These are the experiences of one young Army Pilot in Vietnam from 1969 to 1970. These stories are certainly mirrored in the memories of thousands of crew members and soldiers who served in this conflict.

Check Ride

Check Ride

THOMAS MCGURN

Deeds Publishing | Athens

Copyright © 2020 — Thomas McGurn

ALL RIGHTS RESERVED—No part of this book may be reproduced in any form or by any electronic or mechanical means, including information storage and retrieval systems, without permission in writing from the authors, except by a reviewer who may quote brief passages in a review.

Published by Deeds Publishing in Athens, GA
www.deedspublishing.com

Printed in The United States of America

Cover design by Mark Babcock. Text layout by Matt King.

PAPERBACK ISBN 978-1-950794-06-5
HARDCOVER ISBN 978-1-950794-07-2

Books are available in quantity for promotional or premium use. For information, email info@deedspublishing.com.

First Edition, 2020

10 9 8 7 6 5 4 3 2 1

To all those soldiers, sailors, and airmen who made the supreme sacrifice of their life, I will remember you always.

To all who served in a war of restrictions and the consequences suffered from such exposure. God bless the doctors, nurses, and medics, combat brothers and sisters always.

Contents

Preface	xi
Acknowledgements	xiii
The Inspiration	1

1. Memories	7
2. Check Ride	17
3. That Cat Nearly Killed Me	29
4. 1st Combat Assault as CP	35
5. Ban Me Thuot (Today called Buon Ma Thuot)	41
6. CHARLIE CHARLIE	45
7. Ambush	51
8. LRRP	57
9. TAGI	63
10. The First Time	75
11. Flare Incident	79
12. Thee Bruce	89
13. Pilot in Command (PC)	97
14. Long Day But!	101
15. Sniffer Missions	109
16. USS St. Paul, Last of WWII Gun Cruisers	115
17. Another Night Flare Mission, only as PC	123

18. Lift Mission, Lead Ship	131
19. Why	139
20. Lift Extraction, Mountain Style	147
21. Holy Crap, that's what they look like	153
22. April Fool's Day 1970, Song Mao	159
23. Trusted That Guy	165
24. I Thought They Were Gone	177
25. The Real Times	183
26. Gotta Go	189
27. It Happened So Fast	201
28. Short Timer	223
29. Finishing it out	237
30. Reunion	247
Post War	255
The Special Ones	259
About the Author	265

Preface

How does one arrive at finally documenting those snippets of a soldier's life that lingered over forty plus years? Of the two million seven hundred and ten thousand service personnel who served within the Vietnam conflict, I was one. My cover expresses my basic belief *SOME HAD IT BETTER, SOME HAD IT WORSE.* I place myself in the first category of the fortunate ones. Over five thousand helicopters were lost in this conflict, just machines, a cold avoidance statistic lacking reference to their precious cargo of American soldiers.

"My belief" that history is the long trail to success, imagine, in 1969, I'm flying in a machine that has been refined since its first concept was tested in 1939, only thirty years earlier and nine years before my birth, also grasp that the first powered flight was successfully conducted in 1903, and we were still in the first century of manned flight. What was the advancement during the first one hundred years of the wheel or the bow and arrow, then followed by the tens of centuries to develop and perfect them?

During this long year, I had no comparison to measure my assignment as a combat pilot, just the red neck slogan "get 'er done". My hope is these flying stories will give the reader an everyday flavor of the variety which existed during these times and the path that many had tread. Ranger insertions/extractions, shipborne operations, combat assaults, and the usual WTF! missions conjured by this new generation of mobile warfare warriors and fine-tuned by the needs of the United States Army.

I have written these chronicles with the explanation of certain military terms and equipment incorporated throughout each chapter to enhance the readers understanding and make the stream of events an absorbing process.

Acknowledgements

To my loving wife, Paula Marks McGurn, for her tolerance in my retirement and encouragement to compile such events, and for those interested in one soldier's memories.

To all our daughters, Kerry McGurn DiSalvo, Allyson Marks, Meredith Marks Parrinello, and Samatha Lee. Always giving the strongest of support mixed with love, thank you.

To Charles "Charlie" Morgan, former Air Force B-25 and F-86 pilot then converted into an Army National Guard helicopter pilot, "thank God". Charlie was a senior staff editor for Readers Digest and graciously read my stories and encouraged me to go forward with this book.

To David Paul, author and friend, who wrote *Unearthing a Serial Killer* and invigorated me to publish these true experiences. BTW a graduate of Sacred Heart High School, Yonkers, NY.

To my special and closest friends, I can never thank all of you properly, but I will try in my own way for the reinforcement I received during this writing.

Special thanks to Jim Schueckler and his devotion to our old Unit in his writings, and the inspiration to start this manuscript process.

To Deeds Publishing, Mr. Bob Babcock for his devotion to military related history and documentation of such times. Thank you for that special encouragement in making me believe this book was worthy. I shall never forget your honest but direct quote, "Send me some sample chapters and I will be honest in telling you if your baby is ugly or not."

Special acknowledgement to VHPA (Vietnam Helicopter Pilots Association) for their reference material and Historical documentation of this Helicopter War.

Special recognition to those soldiers who have recorded and preserved many of the pictures of the 192nd Assault Helicopter Companies Area of Operations

Copyright 2003, Ray Sarlin and www.ichiban1.org. All rights reserved. (Copyright policy) rws@173rdAirborne.com

Permission is hereby granted to copy this story to print or on web pages at no charge provided the line below is included: Reprinted from the 1st Bn (Mech) 50th Infantry website http://www.ichiban1.org/ (web sites should make the url a link or may also just link to this page)

Lastly, to Sacred Heart High School, Yonkers, New York, for having that lone pamphlet from the United States Army describing from High School to Flight School. Vaguely reminiscent of General George Smith Patton Jr., iconic statue erected on the hallowed green of West Point where he is eternally facing the old cadet library with a pair of binoculars in his stoic hands, the joke being that this would be the only time he had ever seen it. Well, mine was a similar library experience, however, obviously, I was there once.

General George S Patton Statue, West Point

In the monument's previous position, Patton faced the old Cadet Library. It was often joked that the statue was positioned facing the library with binoculars in the officer's hands so that he might find the building which he neglected to visit as a cadet.

The Inspiration

A FAILED NIGHT RANGER EXTRACTION
PHAN THIET, 1969

(Copyright Jim Schueckler)

Dedicated in memory of CPT Thomas E. "Okie" Campbell, WO1 John P. Wright, SP5 Chubby D. Lowrey, and PFC Ezequiel Torres, Jr.

Flying at night in Vietnam scared the hell out of me. Here in the good old US of A, flying at night is not so dangerous. In most places here, there are thousands of lights on the ground to give references of where the ground and horizon are. Vietnam wasn't like that. The high humidity caused a haze that totally obscured the horizon, and there were no lights on the ground in the mountainous jungle areas. Flying at night in Vietnam was instrument flight. Nobody flies on instruments when close to the ground, but our missions required us to be close to the ground. You can't hover on instruments, because to hover you need very fast information to the brain about the aircraft's attitude and position.

My Peter Pilot was a newby, (Bruce Britton my roommate was Jim's copilot on this mission) about as long as the Mekong River. In Vietnam about a week, he had his "in-country check ride" just the day before. As aircraft commander, I introduced him to supporting C Company of the 75th Rangers, using a Huey H-model slick. We dropped a few Ranger teams in the jungle, took supplies to some of them, brought one team back, and did a few other ash-and-trash missions to firebases and compounds near Phan Thiet. We had thought our work-day was over.

One Ranger team was to stay out in the boonies overnight, but got into a firefight with an undetermined number of VC at about 10 PM. Our gunships and one of our slicks dropping flares went out to support them, and their platoon leader was in an Air Force FAC Bird Dog. We were called to operations: "Go pick up the team. Use McGuire rigs because they can't find a clearing. They are under trees that are forty to fifty feet high."

I had made McGuire rig pickups before -- in daylight. The Rangers on the ground would put on a body-harness with a locking hook at chest level. We would hover above the trees. A Ranger-guy in the back of our Huey would drop long ropes with rings on the end. Very simple: the guys on the ground would connect their hooks to the ring, tell us on the radio or with a thumbs-up that they were ready, and we would pull them straight up until they were clear of the trees. We would fly back home or to a nearby firebase with the men hanging far below the Huey. We had to remember to terminate at a very high hover, then go straight down very slowly. We used to joke that nobody was sure if McGuire was the name of the man who invented the process, or the one who died the first time it was tried. (Children: do not do this at home).

We got airborne within a few minutes and joined the FAC, the gunships, and the flare ship in the air above the embattled Rangers. The area was well illuminated by the flares that our sister-ship was dropping from high above, so I asked the Peter Pilot to fly the approach. We were only a few hundred feet up, with our searchlight on, when tracers came up towards us from several different places. I turned off the lights, grabbed the controls, turned real sharp, and got out of there as fast as I could.

More nervous now, I remained at the controls. We went back, just above the trees, with no lights on except the navigation lights that can only be seen from above. The flares were swinging under their parachutes, making the treetops look like an eerie, rolling, sea. The Ranger on the ground directed me to him by the noise I was making.

The men on the ground were talking on their FM radio, and folks in the air were on UHF, but only the FAC and we were trying to use both.

On the first pass, I was trying to talk with the Rangers on the ground, but the almost continuous air-to-air talk would frequently cover them. I set switches on the intercom units and then told my Peter Pilot: "You talk to the guys in the air, I'll talk to the guys on the ground."

My almost fatal mistake was that I did NOT tell my Peter Pilot that I could no longer hear the air-to-air conversations.

Finally, the man on the ground said we were directly overhead. The Ranger in the back of our Huey dropped the ropes. They tangled. He pulled them back up and tried again. We couldn't see the men on the ground, but the one with the radio could see our silhouette against the glow of the parachute flares, so he gave us directions: "Back up, go to the left, no that's too much, go right..." He had to scream into the radio to overcome the sound of our rotor. I was pouring with sweat and my heart was pounding for what seemed to be an eternity. Finally, the Rangers were connected and gave the signal to pull them up. Because the ropes were over the skids, not at the center of gravity, I had to rise very slowly.

Then the lights went out.

Black. VERY black. All around us.

It seemed as though nothing existed outside of the Huey except darkness.

The slick dropping flares had called out by radio when they had thrown out the third last, then second last, then last, flare. My Peter Pilot didn't know that I couldn't hear that radio. I hadn't heard the warnings.
    There were no more flares. No more light.

From the feeble red glow of the dimmed instruments and the seat of my pants, I could tell that we were spinning. Fast. Spinning like a top, around a bunch of ropes now tangled in the trees. Full opposite pedal did not stop the spinning because we were pivoting around the ropes over the skids.

I couldn't go up because we were tied to the trees, but I couldn't go down for fear of hitting a tree, losing the tail rotor, or otherwise destroying the aircraft.

But I couldn't see ANYTHING outside!

"Okie" Campbell, our gunship platoon leader, saw my navigation lights spinning. He raised the nose of his C-model Huey and punched off a salvo of rockets high into the air. From that few seconds of rocket-flame glare, I was able to stop the spinning. My frantic groping also had finally found the landing light switch. Click. I'd rather get shot at than do that spinning again.

Now able to see the treetops, I reduced power to stop the spinning.

The Ranger in the back of our Huey cut us free of the tangled ropes as he screamed in anguish because he thought his buddies in the darkness below us were dead or soon would be. I felt terrible, wondering how many men I had just killed. As soon as the Ranger said we were free, I pulled maximum power to low-level out of there. After building airspeed, I switched off the landing light and climbed like a bat out of hell.

Within a few minutes, we found out by radio that the Rangers did not fare too badly. They were just a few feet off the ground, and only one was cut from the trees. The Ranger platoon leader had seen it all from the back seat of the FAC plane, and he asked the Rangers on the ground if they would prefer to spend the night where they were. I thanked God when they said they wanted to stay on the ground.

The VC were either killed by our gunships or thought that our mission to pick up the Rangers had been successful. There was no more enemy contact, and the Rangers found a big clearing to be picked up the next morning.

About two weeks later, another one of our (I was the copilot on this second narrative of Jim's story) slicks was trying to extract some Rangers from a clearing on a hill at night. The main rotor hit a tree and the aircraft rolled over; killing one Ranger and injuring several others and the crew. A second gunship team went out to support their recovery. Okie's last radio message said they had inadvertently flown into a cloud; a very bad thing to happen while low in that rugged terrain. We searched all night, finding the wreckage and the bodies of Okie, John, Chubby, and Zeke the next morning. That day our company commander and the Ranger company commander agreed to have no more night extractions. Gunship fire and dropping flares all night if they needed it, but no more night pickups.

Dedicated in memory of CPT Thomas E. "Okie" Campbell, WO1 John P. Wright, SP5 Chubby D. Lowrey, and PFC Ezequiel Torres, Jr.
    Rest in peace, brothers.

Copyright Jim Schueckler, 8219 Parmelee Road, LeRoy, NY 14482

1. Memories

October 2013, forty-four years have passed, memories fade but back then in October 1969, I was still learning. Only recently, my memory was released with a small article authored by Jim Schueckler which mentions the following events in his final paragraphs. Jim was my pilot in command (PC) on this fateful night in the mountains of II Corp, Vietnam. As co-pilots, we were referred to as new guy/newbie/right seat, a position of "do what the PC says and don't ask questions". Jim, in that small memoir, proceeded to tell his story of a rescue mission which was as rewarding and tragic in any two-hour flight period as one could imagine.

From this co-pilot memory, I will tell you that since my arrival in the country of Vietnam, I was in awe of my PC's. How could they, my senior pilots, have acquired so much aviation knowledge being only six months or so ahead of me since their departure from flight school? Where was I on that line when the God of flight was issuing this information, on the job training at its maximum?

Jim Schueckler was my PC and I had already flown several missions with him. He was a true teacher with a quiet demeanor that was not intimidating, you could learn much from this fellow.

On this fateful date in October 1969, I flew with Jim the entire day, only to return late and have our ship and crew placed on a stand-by night alert status. This could be defined as be prepared to be roused to fly any time during this coming night until the next sunrise, "the long day".

After finishing our regular post flight, I proceeded to the mess hall to avoid once again eating food rations out of a can. This facility was closed at this late hour, but pilots and crews were always welcome to be served what scraps may have been left over.

My rendering begins its unfortunate phase here at a dinner table, with our feast consisting of mystery meat, a broiled lump of overcooked animal muscle smothered in a salty gravy to disguise its origin. Usually, this entree was washed down with a dark purple Kool-aid , hopefully with ice. Joining me was John Wright, a fellow copilot who I originally met at the beginning of my Army journey as we both were sworn into the United States Army in the building identified as White Hall Street in NYC.

Both of us were destined to be fellow future army aviators and bonding immediately, together with five other aviation candidates. John and I were in the same flight class 2nd WOC (Warrant Officer Class), known as green hats, which started immediately after our completion of basic training at Ft. Polk, Louisiana, not a pleasant time. Having graduated/ endured that indoctrination, we were immediately shipped out by bus

CHECK RIDE

to our next phase of training, Basic Flight School at Ft. Wolters, Texas, closest town, Mineral Wells.

This location was the birthing center then of all future Army Helicopter Aviators. Now sitting together at dinner one year and six months later, and nine thousand miles from Ft Wolters, Texas, both of us rejoined in the same Assault helicopter company, "Thee" 192nd. John was enthused with a spirit and wanted action, he saw this pathway by volunteering and being quickly assigned early on to the gun platoon, call sign Tiger Sharks. There is a link to someone when you are in the army, some bond, common thread that is an ethereal connection, probably for life, which fosters memories of youth until our aged death.

John and I both sat this evening comfortable with our journey and talked about "old times", imagine old times were maybe twelve plus months. John expressed how he wanted to go north where there was more action. I told John "action has a way of finding us, if it's meant to be." John was assigned to primary guns this night and he would be the co-pilot for Captain Campbell, as I would be the co-pilot for Jim Schueckler.

Night had arrived and was not interrupted by the sporadic incoming enemy mortar or rocket rounds which usually occurred several times during these past months, always at night. Sometime while John and I were finishing our meal, the alert-horn sounded for primary ships to launch on a mission unidentified to most. The entire Company knew what this meant, men going into the unknown night for someone in contact (**CONTACT**: a firefight, being in **contact** with the enemy) that absolutely could not wait for daylight. John looked at me and said, "have to go," which I acknowledge with a "be safe." Who knew what was about to happen? No hunch, or hair standing up on the back of your neck, just two army buddies expecting to meet again.

Two slicks (troop carrying helicopters armed with two M60 machine guns) and two gunships (helicopters specifically armed with mini guns and rockets to cover troop carrying helicopters) into the blackness with rising mountains four thousand feet in height only eight miles to the west. Perilously close when flying over one hundred MPH, business as usual.

Now as standby-alert, Jim, myself, and the crew progressed to primary alert aircraft, never expecting to launch. It might have been about two hours when the alert horn sounded again and I would be flying with Jim into that same darkness toward those same mountains, learning our mission while en-route, something has gone wrong.

We take off from Phan Thiet, AKA, LZ (also known as, Landing Zone) Betty, westward toward the mountain jungles. I'm directed by Jim to climb and hold a heading (*The direction in which an aircraft is pointing*). We lowered our interior lights to adjust and improve our night vision. Jim now took over. I felt confident with him, as I stated, he was a teacher, someone that was showing me how to survive when these men would go home one day within these next six months.

How often will you read in novels how pilots and soldiers strained their eyes looking into the dark? The sky was so black, no stars, no horizon as we flew at ninety knots on maps, time, heading, speed, altitude, and instincts. This would be considered the dark ages of Army aviation, no GPS (global positioning system), no NVG (night vision goggles), no heads-up displays (*Display of instrument readings in an aircraft, typically through being projected onto the windshield or visor*). I was totally overwhelmed and intently listening to Jim as he now took the controls while directing the mission.

Four helicopters, two slicks (troop transports) and two gunships comprised the first team which launched and, unfortunately, one had crashed during the troop extraction, resulting in the death of one member from the ranger team. The gunships never land; their assignment is to orbit and provide armed support for the slick aircraft while they are exposed on the ground in enemy territory. Our mission was to now recover members of the first alert team whose chopper blade had caught the slope of the mountain side and then rolled down the LZ (Landing Zone) turning into a tossing ball of aircraft parts aided by the centrifugal force from the turbine powered rotor head.

They had been in the process of rescuing a LRRP (long range recon patrol) team who most certainly had to have been in dire danger to request such a hazardous mission this night. This was one of the 192nd Companies primary missions, to support ranger operations. If Army pilots and crews bring you in, we will endeavor to get you out, no matter what the circumstances, it's our vow to the grunts/brothers.

Jim started in with reducing power and then directing me to now turn on the landing light. This was a little disturbing to me, instantly seeing the outline of those huge jungle trees and suddenly realizing we were below the height of these mountains; a technique in flight school we were instructed to avoid! Especially at night.

Also, some relief at the reverse sensation of this vision, for now I could see the thick jungle vegetation on the hillside, giving us some welcome visual reference for positioning. Jim, using all his acquired skills, positioned our aircraft parallel to the slope. I was seated on the down slope side, still following Jim's directions, with our landing light and updates on available power. I could clearly see the ghostly hulk of the wrecked Huey crumpled on its back, with no blades, up against the tree line.

This experience flooded my mind as my eyes conveyed what I perceived as an unbelievable situation and how did I get here. We cautiously extracted our people, knowing our uphill blade position was within a few feet from contacting that same steep slope which snared the first rescue mission whose skeleton metallic remains of this former helicopter was still in my sight on our right side. Finally, all pacs (passengers) are safely onboard as we pull in power, climbing to clear the mountain, and head for home.

Immeasurable relief was my feeling to be on our way back to our base, my first night mission was a success and we were alive. I was never worried, more comparable to concerned, knowing such future missions may be forthcoming. I had total faith in Jim's ability this night, with my contribution only being an exceptionally minor input as the copilot, but the learning lesson was tattooed into my brain. Months later as PC, I would be tasked to use such skills myself and, believe me, I thank all the PC's

I have ever flown with for their attention to training. For that day did come when my crew would be responsible for a rescue of a LRRP team at night, using a rope ladder and maintaining the position of our ship by placing the left side chin bubble on the tree top to maintain a constant position and I hope I passed on needed experience to my co-pilot, Lt. Hoss.

Inbound now to Phan Thiet, dropping off pacs, and refueling when we learned that we have lost two ships on this October night. John Wright's gunship is missing with one contact since they first left the base, transmitting that they were in some ground fog while climbing through a mountain pass. We immediately refueled and headed back into those black mountains to search for our comrades, brothers, family, and soldiers all. We flew the ridges at tree top level, knowing the gunships, being heavier, climb slower and always face more hazards. During one run while Jim is near the top of a saddle (low point on a ridge between two summits) I spotted a fire directly below.

Jim now positioned our ship, hovering so I can look down on the right side. Flames were clearly visible, burning in an irregular pattern as seen through hundred-foot jungle trees below in the dark as we continued to hover overhead, but we couldn't confirm anything through the dense canopy of these jungle trees and generated smoke. Often during the dry season there would be scattered fires in the countryside, not unusual. We continued searching until just before dawn when we received instructions by our company command to return to normal mission status.

The war went on and our primary purpose was supporting numerous units on a daily basis, our living soldiers needed us. Last, I was flying the next morning again as co-pilot, knowing my friend John was gone as I reflected on the casualness of our last meal together.

The gunship would be found later, as I understood, by a medevac searching several days after the crash. In Jim's memory, he mentioned it was found earlier, but that's why we shared some of these tragic memories. I was told if the ship had caught fire, the vegetation in the area could eventually turn brown and you may be able to observe this different color tone. Some days later, while flying over the area of the suspected crash location, they were discovered, again learning lessons of war.

Our company lost four good men that night, something I wasn't prepared for, and only recently had been updated that a young Army Ranger also lost his life during that crash of the extraction ship. At the time of that event, it aged this twenty-one-year-old into reality, as memories of John flashed, knowing he was gone.

I had left NYC from White Hall Street with seven future army aviation candidates in May of 1968. Two would not make it through flight school, choosing to drop out. Two would be killed in action in Vietnam. One would be killed later at home by police, while suffering PTSD. This I learned from his cousin, Jimmy Coan, who was now a pilot in my guard unit and would go on to be a NYPD Inspector in charge of NYPD Aviation. Jim Dixon and I would go on with life.

After reading Jim Schueckler's account of that night, I was not offended being referenced to as a co-pilot, but received a collective jolt to my memory, reminding me of an essential learning process that helped all copilots to survive later. Hopefully, I passed on skills to those while I was a PC.

Most of all, when I'm enjoying some special time in my life, I try and remember those who didn't get the chances I did, and those whom always stayed forever young, but not forgotten while I breathe. As per Jim's request, this is dedicated in memory of CPT Thomas E. "Okie" Campbell, WO1 John P. Wright, SP5 Chubby D. Lowrey, and PFC Ezequiel Torres, Jr.

— CW4 Tom McGurn
Retired

2. Check Ride

May 9, 1969, I graduated Army Flight School and received my Army Aviator wings, one of the proudest times in my life. I gave that first set of silver wings to my mother, who with my father had travelled down to Ft Rucker, Alabama to be there for me. What was so meaningful was I always thought I let my parent's aspirations for me down, always drifting through school with no direction, no goals. With my completion of this very difficult military flight school, I was now graduating as an officer and pilot. I could see my Mom and Dad's worries fade, finally seeing true pride in their expressions. Good times when those you love and who loved you can be given a gift which recognizes their efforts as parents.

I would leave my home, that familiar piece of my earth, around May 30, 1969, to begin this odyssey which I knew was coming. Once you voluntarily enlisted to go to the United States Army Flight School, your final destination was always going to be Vietnam. A deal with the devil, and a mutual contract with Uncle Sam, for the reward of flying. I would leave from JFK Airport as my parents who drove me down then said their goodbye, as my best childhood thoughts with them surfaced, remembering walking as a small child, holding both their hands, feeling that utter sense of protection which only a mother and father can project as they initiated my first taste of flight, as my little feet left the earth with both parents enhancing that moment with a mandatory "weeeee". I often pondered if their thoughts conjured that I may not return from this sojourn, and how they both possibly bore such a consequence on their drive home to Yonkers and into this next protracted year of what if's, myself instantaneously questioning would I see my loving parents again.

As I boarded the plane I felt alone, unaccompanied, but not homesick, I went through that emotion during Basic training and was well beyond that emotion, now realizing home is always fixed, it's us that move. On my long flight to Seattle, I tried to imagine my path as I moved forward. Landing in Seattle and then at Ft Lewis, Washington, the military point of my final departure from my country, my home, and this familiar center of my universe these past twenty-one years. One more giant leap until feet down in another world.

All processed at Ft. Lewis within twenty-four hours and scheduled to leave on a DC8 Overseas aircraft. I was twenty-one years old, something I may mention too much, and with another young pilot we were the only two officers on this plane's manifest of one hundred and sixty passengers. Our route was north to Anchorage, Alaska, for refuel, then onto Tokyo, disembarking just long enough to see Mt. Fuji in the distance, not some poster picture, as we once again refueled with the final destination Cam Ranh Bay, Vietnam. Eighteen hours from start to arrival at Cam Ranh Bay, a huge air force base on the coast near the City of Nha Trang, VIETNAM. Where hundreds of thousands of soldiers would arrive, but not all would leave in seats.

Off the plane into the tropical heat and a roster was called with a shout of follow me from some guide to our next location. We had been split into several groups. Thank God, we were young and in shape. We walked everywhere carrying our bags and duffle bags in this unbelievable heat during the night, not love'n this. You could hear the running CH

47, a large two rotor Army chopper which smelled of burning jet fuel, as we were told to board and strap in. The passage commenced.

Off into this humid dark night like mushrooms, no information, just ear bleeding noise, and the familiar waddle associated with this aircraft's flight characteristic accompanied with that strong odor of burning jet fuel being inhaled on empty stomachs. Just as we began to settle in, we were on approach to land. You're kidding (all that for) a ten-minute flight?

We landed at a compound in Nha Trang. Very late at night, now finally we had some kind of human contact to explain what was going to happen. The soldier explained we would be taken to a barracks to sleep, then moved in the morning to our assigned unit's headquarters for eventual permanent placement.

Waking on my very first day in a war zone after an uneasy night's slumber in a bed which has had more customers than a Mexican whore house, I'm was up taking in the first foreign daylight, diverse with humidity and unfamiliar odors of this Asian City. I quickly realized we were barracked right up against a barbwire fence in the City of Nha Trang, you could have poked me with a stick from the other side. It was not the secure perimeter I expected.

First, I viewed a few emaciated chickens pecking a hard-packed pathway and then a small group of children smiling with timeworn clothes. Later that day, I was placed on a UH-1 helicopter from the 92nd Assault Helicopter Company stationed at Dong Ba Thin and flown to that location which was the headquarters of the 10th Combat Aviation Battalion of the 1st Aviation Brigade. I would be processed at this HQs and receive my final posting to the 192nd Assault Helicopter Company, located at Phan Thiet, call signed Landing Zone Betty, a lone outpost on the coast.

The next evening, I was placed on a 192nd ship en route to my new home. I was clueless and felt like a package in the mail. Flying with doors open, sitting right on the edge of the seat, strapped in taking this journey through the mountains and rice paddies below. We passed Phan Rang, a huge Air force base, and then flew through a steep mountain pass, learning later the name is the Phan Rang Pass.

Now seeing the expansive South China Sea on my left and the lush dark green jungled mountains on my right, it was nothing like the training fields of Fort Wolters or Fort Rucker. So it was official, training was definitely over, "I'm in it now". This journey lasted about two hours, watching as the sun began to set behind the mountains. Approaching Phan Thiet, you first saw this exotic city on the coast divided by a river and just beyond, on the plateau, the crew chief pointed to my first view of LZ Betty.

LZ Betty

CHECK RIDE

Enlisted Row

His face was smiling as he was the first individual to introduce the new guy to a tent city where all the buildings were surrounded by sandbags. LZ Betty was genuinely a desolate place with tents and some new metal roof buildings with unpainted wood sides. Running the entire length of the LZ was a PSP (perforated steel planking) airstrip enclosed by a substantial perimeter and what I expected was lots of gun towers with a cleaned area in front for enhanced fields of fire.

One ominous and prominent manmade terrain feature which was the entire west side of the base, just past the perimeter wire, was a graveyard, for those who believe in such portents.

Graveyard

Upon landing, I met Larry Hupe, a pilot in my new unit who was assigned night duty officer that day. Of course, this assignment came after he had been flying all day. Larry graciously escorted me to his room where I could crash on his bed that first night, soldiers taking care of soldiers.

Larry, some months after my arrival at LZ Betty, was flying gun ships and unintentionally flew through a mist of Agent Orange being spayed by Bruce Britton, my roommate, from his appropriately rigged UH-1H over the Le Hong Phong Forest. I would later meet Larry after I left Vietnam, at Ft Rucker, Alabama, where he was assigned as a tactics flight instructor while I was assigned as a Contact flight instructor, now training pilots in turbine powered helicopters including all emergency procedures associated with these new machines. Larry had developed a serious physical condition caused by a cyst outside the lining of his lung while stationed at Ft Rucker, which would have to be operated on to be removed.

The doctors nearly cut him in half, the scar went from his mid-back to his mid-chest. I can only hope Larry received some form of compensation for this permanent scar and suffering. At the time, the army told

CHECK RIDE

him the cyst was outside the lung and not caused by Agent Orange. "Sure, I buy that one."

About two days had passed, I had a room assignment. I was issued flight gear which included a chicken plate (armored chest plate for flying) survival vest with PRC90 (personal survival radio) and blood chit (silk scarf with multiple local languages stating this is an American Pilot if you assist him we will reward you) and a Smith and Wesson six shot 38 caliber revolver with a box of ammunition.

Now it was time to start flying, but first a check ride. All through flight school, from beginning to the very end, there were check rides and each one was critical toward your completion of flight school. Every pilot always dreaded these spot checks, it could mean you were set back, or it could mean you were dropped out of flight school and going to the infantry, but failure was what it meant to me.

This check ride was a little different, I was already a pilot and I didn't have to be concerned with losing my silver wings, but I did now have to be concerned with being accepted after this instructor pilot's flight and evaluation report.

I met the Battalion instructor pilot, CW2 Rich Arann. He had made this special trip to LZ Betty to give me my orientation ride which would begin my combat flying career. I had forgotten his name but fortunately I was able to locate it on the Vietnam Helicopter Pilots web page and have included this sad chapter below. CWO Arann taught me so much in this two-hour flight, if I could only thank him for his insights which have guided me throughout my life. He started by relaxing me, letting me know this check was not pass or fail but an introduction to what I may face in this coming year.

We talked the entire two hours while completing autorotation's, run on landings, hydraulic off landings, emergency governor operations, radio procedures, emergency operations and, most importantly, how to deal with the next three hundred and sixty days. He talked about himself and what he wanted to accomplish when he returned home. This was his second tour and he loved flying in the Army. It was so comfortable on this ride, nothing like I expected, a true encouragement.

Finishing the flight, I reported to the 1st Platoon commander for my first true assignment. All the pilots knew this Instructor and he was held in high esteem, which I agreed was justified. Sadly, this young instructor would be murdered in Dong Ba Thin on June 24, 1969, two weeks after my check ride. The Chief was sleeping when a private mistakenly fragged (deliberately kill an unpopular senior officer, typically with a hand grenade) his room with a claymore mine, killing this fine young man instantly. The private had fragged the wrong room. He had deliberately intended to kill his commander who had issued him an Article Fifteen, which permits commanders to administratively discipline troops without a court-martial, but this horrible intentional mistake killed an innocent man who was so well respected. All his dreams that we discussed on that two-hour check stopped forever by some miscreant who felt abused by his commander.

I will never forget this man's kindness and incorporate this same approach of teaching when I would also be an instructor pilot training new pilots at Ft Rucker as tribute to Mr. Arann.

Moving forward thirty-five years, I was assigned to the 42nd Aviation Brigade and living at FOB (Forward Operation Base) Danger, Tikrit, Iraq, when another similar tragedy happen on June 7, 2005. Another disgruntled sergeant would frag his commanders' room with the same weapon, a claymore mine, killing both officers inside. Their names were Captain Phillip Esposito and 1st Lt. Louis Allen. Two times I have been deployed to combat zones and during both of these tours, separated by decades, good soldiers were killed, in the same month, by the same weapon from hostile action within. I live my life with such remembrances and always hope their families know some of us will never forget their sacrifice.

Claymore Mine, A Horrific Weapon

ARANN RICHARD MAXWELL

```
Name: CW2 Richard Maxwell Arann
Status: Killed In Action from an incident on
06/24/1969 while performing the duty of Pilot.
```

    Age at death: 28.0
    Date of Birth: 07/02/1941
    Home City: Norfolk, VA
    Service: AV branch of the reserve component of the
    U.S. Army.
    Unit: 192 AHC, 10 CAB
    Major organization: 1st Aviation Brigade
    Flight class: 66-13
    Service: AV branch of the U.S. Army.
    The Wall location: 21W-005
    Short Summary: Killed by mistake by a claymore mine
    set to kill the company commander.
    Service number: W3154117
    Country: South Vietnam
    MOS: 100B = Utility/Observation Helicopter Pilot
    Primary cause: Ground Casualty
    Compliment cause: mines
    Started Tour: 04/13/1969
    "Official" listing: ground casualty
    Length of service: 10
    Location: Tuyen Duc Province II Corps.

Additional information about this casualty:

Chief Arann was on his second tour in Vietnam when he died. On the night of 24 Jun 69, PVT William E. Sutton, who was angry at his CO (CPT Angeli) because he received an Article 15 for smoking pot, detonated a claymore outside of the billet where he thought Angeli would be sleeping. He killed Arann by mistake. Sutton was court-martialed and sentenced to life imprisonment. His sentence was cut to thirty years and he will be released from the Oklahoma Transfer Center next month. From George Lepre, April 1999. On 7 Aug 99, the last

Vietnam-era fragger will leave prison. Ex-PVT William E. Sutton will be released from the federal prison in Yazoo City, MS (Although most military offenders serve their sentences at Leavenworth, during the early 1970s, a number of particularly incorrigible inmates were sent out into the regular federal system). Sutton, a signal wireman in HHC, 10th Combat Aviation Bn, 1st Aviation Bde, had received an Article 15 for smoking marijuana from his company commander, CPT Robert Angeli. On the night of 23-24 June 1969, Sutton placed a claymore outside the CO's quarters and detonated it. However, he was so stoned when he did it that he placed it in the wrong part of the building and he killed helicopter pilot CW2 Richard M. Arann by mistake. He then ran to a second mine he had set up to kill the company first sergeant. This mine too was set up in the wrong place and it seriously wounded the battalion sergeant major, Grant McBee. CW2 Arann was an outstanding pilot and was on his second tour in Vietnam. Sutton proved to be a disciplinary problem while he was at Leavenworth. He was charged with failing to stand count, disrespect to a guard, homosexual assault, and a number of other offenses. As a result, he was quickly sent out to the regular federal system. He was briefly paroled in 1980 but quickly re-arrested on a violation of his conditional release (I presume he burned a piss test). George Lepre Co. A, 2d Bn., 4th Inf. (Warriors)

* * *

```
Reason: other accident
Casualty type: Non-hostile—died of other causes
married male U.S. citizen
Race: Caucasian
Religion: Methodist (Evangelical United Brethren)
The following information secondary, but may help in
explaining this incident.
```

THOMAS MCGURN

```
Category of casualty as defined by the Army: non-bat-
tle dead Category of personnel: active duty Army
Military class: warrant officer
```

This record was last updated on 04/26/2002
Date posted on this site: 01/22/2013

Copyright © 1998—2012 Vietnam Helicopter Pilots Association Vietnam Helicopter Pilots Association

3. That Cat Nearly Killed Me

FINAL THOUGHTS BEFORE DEATH?

Originally this memory writing exercise began at the request of my family. They so often heard bits and bobs of stories, but never received the whole, true picture according to CW4 Tom McGurn. Some stories I glossed over, wishing to spare them the gruesome details. And some, to be quite honest, stirred up more emotion than I was willing to handle at the time. So why now, you ask? To be frank, at my age (sixty-six years young), I think my family believes it is now or never! Perhaps my stories will provide authentication for those great grandchildren who desire to sell my ancient military paraphernalia on the Antiques Road show. Hey, if I can make them a buck from the afterlife, why not?

This story is another co-pilot adventure, one that found a young me flying with an unnamed PC (call sign "the" Duck). Our mission: a MACV (Military Assistance Command Vietnam) bus run. These missions were the everyday/ daily dedicated trips that covered the entire area of operation for which the 192nd Assault Helicopter Company was responsible. We would fly from LZ Betty to several towns and villages, transporting military and civilians so they could avoid the many hazards of travelling on "Highway One" where ambushes were all too frequent. Obviously, this was the safest mode of transportation for our soldiers and those selected civilians. Clearly this accelerated ones' familiarization of our area of operation, while enhancing map reading skills mixed with terrain recognition, and achieved the much needed practice of multiple landings and takeoffs into local villages. The mission was usually assigned to new PC's and co-pilots. Someone really thought this out because it certainly was an effective apprentice tool, which I would appreciate profoundly as this long year continued.

The day began normally; we had just departed the village of Song Mao with a mixed group of civilians and soldiers (as passengers) now flying eastbound to another small village near the coast. I had flown with this PC on several occasions and he had a hot dog approach to flying ("hot dog" used here is aviation slang for doing something special to capture attention). Look, all pilots love their machine and its capabilities, flying is like a drug, and most love a good roller coaster, and flying Army helicopters. Well, it is the ultimate coaster, for those car buffs a screaming airborne red Ferrari but painted olive drab. We were now flying/cruising along at about one thousand feet above the forest below with our airspeed just over one hundred knots approximately (115MPH). Normal altitude would be fifteen hundred feet to avoid small arms ground fire.

My PC "thee Duck" unexpectedly and without any crew notification chose to frighten the passengers on board, for reasons only known to him. "Thee Duck" performed an unannounced out of envelope maneuver (maneuvering beyond the designed aircraft capabilities, aka—no bueno). Now, the PC is the ultimate and only authority on this ship with the copilot and crew along for the ride as his pawns. Over my forty-year career in Army Aviation, emphasis on crew coordination would not be incorporated until many years after this event, finally ensuring a more consistent flight profile and possibly avoiding what was about to take place, a no notice, no discussion, occurrence that caused life threatening concerns.

The PC now whips the helicopter into an extreme left turn, steeply banking while applying full left pedal, the forces are felt instantaneously as I immediately reach for the overhead handle hold. I quickly glance left and see the shock on his face, that intense "OH fuck!" expression and I think: this ass has just killed us all. He battled to regain control. With my certain death in the forefront of my mind, my honest to God only thought was, "I hope this doesn't hurt". The aircraft is now one hundred and eighty degrees out from our previous heading, wobbling back and forth, ass end down due to our forward momentum, being suddenly reversed and the aircraft now unable to dissipate those aerodynamic forces. Imagine going over one hundred miles per hour in your car when the front is now facing the back, not adding the additional factor of altitude.

Going backward, we sank toward impact in the Le Hong Phong forest below. The helicopter shuddered violently, I felt it may break up any second and was surprised the tail boom section was not torn from the fuselage. In my entire flying career, I have never experienced anything this intense in an aircraft. As the recovery was being attempted, I was totally helpless: trapped in someone else's mistake. During this event, you could hear the screams of the civilians/paxs (passengers) through our helmets even with all the aircraft doors open, if only they knew this guy lost it deliberately.

By some miracle (one of many, for me, my opinion), we recovered about one hundred feet above the forest floor. I looked at my aircraft commander's face, which has drained to that death white color. Upon recovery, I immediately went to private intercom to tell the PC if he ever does that again I'll FN kill him. You could see the PC scared himself and was now trying to recover some composure, not a characteristic you need on any mission.

We finished that flying day of about eight hours, and I made my way purposefully (ok, more like pissed off) to the 1st Platoon Commander to tell him I will never fly with this guy again. I did this despite fearing all the 1st Platoon PCs' reactions to a copilot's accusation, possibly marking me forever and making the long year ahead uncomfortable. However, they all recognized I was truthful, and none had issues with my reaction to this event. This was not this PC's first blemish: he was getting a reputation, and his future would be restricted to less involved missions until he left Vietnam.

I will include one earlier mission I flew with this PC and promise to mention him no mas (using my limited Spanish vocabulary to accentuate never again). We were conducting a resupply log (logistics) mission out in the flats (not mountains but low lands) just west of Titty Mountain/Whiskey Mountain, an orientation feature from Mother Nature, a single mound of earth that went straight up over twelve hundred feet and was capped with a small detachment of soldiers who maintained a radio communication post. From here they observed most of our area of operations below and from this location had a spectacular 360 degrees of beautiful, if war torn, panoramic view. This outpost had an incredibly small pad, barely large enough to accommodate the helicopter landing skids at its peak, designed for resupply; a true challenge. Over the years, there were helicopter wrecks near the top of this pad, a testimony of the difficulty of pinnacle approaches with a full load. During my tour, I would make numerous approaches and departures to this location and one's experience is accelerated when life is involved.

Back to my story. We were on short final to a dry rice paddy to drop supplies to our troops and we had just planted our skids on the ground when we started to take some small arms ground fire. The crew chief/gunner wasted little time in emptying the cargo floor as we then rushed to pull in power and clear the area. The PC (call sign Duck) was still in communication with our troops on the ground and volunteered our aerial observation to attempt to find the enemy. We spotted them in a dry stream bed as they fled toward an overgrown patch of grass and trees, scrambling between the bushes. As we continued to orbit overhead, the PC updated the ground troops who were in the process of positioning several APC's (armored personnel carriers) to encircle the area that the Viet Cong had fled to for false safety, (Viet Cong (VC), in full Vietnam Cong San, English, Vietnamese Communists, the guerrilla force that, with the support of the North Vietnamese Army, fought against South Vietnam).

Orbiting directly overhead, we observed an APC with a flame thrower we called "zippo" engulf this area with a steady black oily stream of fire. Ground troops inspecting this area after the inferno later found a small mud bunker and located two teenage Viet Cong (a male and a female) both dead, and probably died of suffocation from lack of oxygen. Inside the bunker, our troops recovered several small arms weapons and grenades, not the toys teenagers should have. The PC was stoked that we took part in the action and later painted a little red man with a Chinese hat on his aircraft to commemorate his participation in their death. I had a different view. Look, these people, these kids, were shooting at us and I don't regret their fate. But I never celebrated it either. Somewhere I have a picture of me, now as a PC, flying that ship with that symbol still painted on the side. This would be the second time I had taken fire during missions which I participated in. The third—well, that was a survival miracle.

4. 1st Combat Assault as CP

My first combat assault mission came about one week after I had been flying the bus runs. Naturally I was excited, this was it, and the real deal, not some simulation we conducted during training flights at Ft Rucker. The mission that day was to insert a Company of ground troops into the southern part of the Le Hong Phong forest. This forest was a large plain with very overgrown dense low trees and grass. Now appreciate that the UH-1H helicopter normally carries six fully loaded combat troops with an estimated weight of two hundred forty pounds each, all sitting on the cargo floor, not strapped in. The Army is remarkably accurate in their manuals, especially concerning aviation practices when weight and balance are essential. Some of these soldiers may have weighed in at one hundred and fifty pounds but add water, food, ammunition, overnight gear, and a whole list of essentials learned from combat expedience now stuffed into their rucksack and they would always weigh considerably more.

Think of it like taking a family of six to the beach, shade umbrellas, picnic basket, beach chairs, toys, and how to tailor all this equipment to fit in your car. There is no turning around because you forgot something.

Engine power is king, and when talking turbine engine power, we always monitored torque on this machine, which at fifty pounds indicated displayed on your analog gage is maximum with that familiar red line of do not exceed in your face. After the torque needle indication pegged at its red line, the engine and rotor RPM gauge would start to bleed off, the first indication that your ship was over gross weight. What this really means is the RPM gauge is dropping below the green operational limit, you're losing power, and you're losing lift, contributing to problems taking off or while landing. What also affected the aircrafts lifting capacity was ambient temperatures, particularly on very hot days. Hot air rises, making it less dense, and Vietnam was always hot. So, the majority of lift missions were conducted in the early morning AM hours to take advantage of the cooler temperatures.

The normal mission process worked something like this, the ground Brigade or Battalion commander received a directive to conduct an operation from his higher command. At that point, determined the needs of the mission and assigned his company commanders to prepare to execute the assignment, after which they then determined if aviation assets would be needed. Next, these commanders forwarded a written mission request to the Aviation Brigade who then assigned the Assault Helicopter Company for that particular area of coverage to coordinate a plan for the air assault movement in conjunction with the ground troops.

There were multiple types of missions and planning for some was very complex. Most missions were on a daily basis with fixed planning in place. More urgent missions may involve minor planning to more complex solutions to aid in their completion if necessary, depending on varied situations. Always expect the unexpected, and literally plan on the fly.

This day's air assault was to be completed with six slicks each from 1st and 2nd Platoons, a total of twelve helicopters, accompanied by three gunships from the gun Platoon. The gunships would provide covering fire during the initial landing, again looking forward to my first real life air assault. Slick is the name for troop carrying UH-IH helicopters; each slick had a crew of four and two mounted M60 machine guns, one on each side for defense and suppression of enemy fire, a wicked weapon. Aircraft crews would also have personal weapons, M16s rifles for the gunner and crew chief, and for the pilots a six shot 38 caliber useless S&W revolver with ball ammunition. I would subsequently be informed by all my Company PCs to acquire a more substantial personal defense weapon.

Most missions started with a before sunrise crew brief then followed by preflight and multiple radio checks. Timing had been briefed and established prior to load pickup. There would be two separate flights of six helicopters, approximately five minutes apart, and each flight would make two turns or trips/ sorties. The PZ/starting point (pickup zone) would be from our base at LZ Betty. The first flight formed up into trail formation, rotors turning in unison, revealing their awesome power as the troops were lined up conspicuously fully loaded, poor lads hunched over with massive rucksacks and weapons waiting to be transported to their unknown reception, good or bad.

All onboard, three troops on each side, no seat belts here, once fully loaded the crew chief and gunner gave us a go. The lead aircraft announced five seconds to pitch pull. That notification is designed as a get ready call. The flight took off, all six-aircraft straining to gain lift, hampered with full fuel and loaded with combat troops. Our flight was a short one, maybe fifteen minutes as we changed formation en route, commanded by the flight lead as we formed up in a staggered right. Flight formations vary from number of aircraft to size of LZ and door gunner coverage effectiveness for enemy suppression. Communication was primarily conducted on two radios, FM for ground troops and UHF for aircraft.

The radio chatter was nonstop as the gunships arrived to prep the landing zone. Cross checking the maps, I could now see the landing zone as we turned to our approach heading, a large very low brush area surrounded by the low forest trees, we were the number two ship on the right side. The command came for guns red (I may have to check this verbiage, memory doubt, but something like this) so in this particular formation only the outboard gunners fired. The inboard gunners remained cold to ensure no accidental rounds could ricochet into one of our own aircraft.

On short final, our flight had lined up about one half mile before touchdown. The LZ was a large one, located in the Le Hong Phong Forest a substantial uninhabited area which extended north maybe thirty miles and was bordered on the east by the South China sea and on the north by Highway One. The landing zone was filled with high grass and contoured with a slight right to left sloped area, surrounded by that very dense low vegetation of the Forest. Our gunships had joined us, paralleling our inbound course slightly higher and just forward of the flight, they were starting their suppression run. This was it; I was trying to imagine what was going to take place in the next two minutes while completing all the tasks that the PC requested of me. Gunships doing their job let loose, flying forward keeping their speed as we decelerated to land, their rockets and minigun fire raking the tree lines, you could feel the concussion of the five to seven-pound explosive warheads from these tubes of death as they hit the ground fifty to one hundred yards away, closer than I had ever anticipated. The lead ship gave the command for right side door gunners to go hot, followed with a cacophony of chattering machine guns. I could see the red tracers of our bullets rip into the tree line and disturb the earth. All the radios buzzed as each ship touched down with very little ground run, soldiers yelling, pumping themselves up to get going. You never really will know if the enemy is in the tree line until you get there, so going in hot is the accepted practice, which I wholly agree with.

The troops quickly ran several steps from the helicopters and took their fighting position as I think how fortunate I was to become a pilot and only have to deliver these brave pawns to this wasteland. Lead called for lift off as all six ships sprang into the air, unburdened by the excessive weight of manpower. Now on the ground, thirty-six men per formation, a total of four formations, and at completion of this mission approximately one hundred and twenty of America's finest would be on the hunt for Charlie, approximately one company. God Bless them all. It would be a sad day when I would be picking up wounded and dead soldiers on some future missions.

"Viet Cong" was commonly shortened to "VC", which in the phonetic alphabet is pronounced "Victor-Charlie", which became, "Charlie."

We returned to LZ Betty for fuel and continued on with our secondary mission, mostly resupply of other units scattered throughout the AO (area of operation). Fortunately, there was no enemy ground fire received on this, my first air assault mission, but there was a long, long, year ahead and as fast as it went by, time seemed to move so slow then.

Le Hong Phong Forest

5. Ban Me Thuot (Today called Buon Ma Thuot)

We sometimes flew outside our "Area of Operations" and on this occasion, it was an all-day excursion plus which took us north. We travelled up the coast through the Phan Rang pass to Nha Trang, a beautiful city on the South China Sea with a large white Buddha on a prominent hill, my first detailed look since my arrival. After refueling, we continued on our flight northwest toward Ban Me Thuot. At that time, Ban Me Thuot was a small city deep in the Central Highlands and manned by one of our sister Assault Helicopter Companies, the 155th. Ban Me Thuot was one of those special cities marked by the North Vietnamese as an entry point and was often under attack as it was situated only twenty-five to thirty miles from the Cambodian border. Thank God, I didn't end up there, it was danger close to the Cambodian border and the Ho Chi Minh trail, giving the enemy access to a readily easy target. The terrain around this City was mountainous, with triple canopy jungle and a different climate of tropical rain and clouds often hampering aviation operations. It's amusing to realize that you feel fortunate to be in a war at a different location knowing now it's only one's own perception or a false sense of security.

Small world alert...I would later meet CWO Bruce McInnes another 2nd WOC sorority graduate (of course tongue in cheek reference since we all chatted like young girls in flight school) years later in my NY Guard Unit. Bruce had been assigned to the 155th in Ban Me Thout during my Vietnam tour and would become a multiple Brownstone Building owner in Brooklyn, NY, during the late 1990s as the millennial occupation was taking place. "Good for Him." SOME HAD IT BETTER.

I cannot remember who my PC was on that date. Despite having flown half this day to arrive at what must be the northern arch of our flying zone, I recall my impression was that no one really knew where we were.

We arrived at Ban Me Thuot, dropped our cargo and paxs (passengers), refueled and prepared to start our lengthy journey back to our home base at Phan Thiet/LZ Betty via the City of Dalat. Our new route would be a direct flight to Dalat, located in the central mountains, a flight of about two hours directly over mountains and triple canopy jungle, no roads, no villages, no force landing areas, and no good flight following or communications. This area is not without its own history with regard to helicopters.

Years after this first experience, I would become good friends with Rocky Muldoon, now in my NY Guard Unit. Rocky had been stationed in Ban Me Thuot a year and a half before I would be serving in Vietnam, Small world again. I had first seen Rocky while he was a primary helicopter instructor at Demsy Field, Ft Wolters, Texas in 1968. Rocky had the most incredible story of survival, one of those GOD moments. Rocky was also a co-pilot and flying the same route from Ban Me Thuot to DaLat only at night when they lost their engine and auto rotated (**Autorotation** is when the main rotor system of a helicopter turns by air moving up through the rotor, rather than engine power driving the rotor) into those jungle mountains and obviously survived. Can you just use your imagination? Loss of engine power at night, you now have to maintain airspeed and reduce your collective which controls the angle of attack of your rotor blades to maintain rotor RPM, while dropping like a rock into the darkness, fumbling for your landing light switch to warily see what's approaching from below, that immediate feeling of panic as they high speed sink toward the jungle. Their crew miraculously survived and were located and pulled to safety after crashing through jungle canopy with trees fifty to one hundred feet in height at night. Miracle maybe, but it was good luck to fly with Rocky. Less fortunately, while assigned to the 192nd my Company too lost an entire crew of four from the 2nd platoon and three additional passengers during November 1969 on a mission over these highlands en route to Dalat during a daytime mission. They were found three years later, with seven fatalities, due to bad weather in these rugged mountains.

While never having known of Rocky's event, these are the concerns of our crew on this daytime flight; you constantly look forward and down, viewing the most rugged terrain and steep mountain passes thinking, *"Who would know what happened to them,* **besides Rocky***"?* Now keep in mind we have always been told that the engines of the Bell helicopter are 99.9% reliable. Every pilot always fixates on that .1%.

We arrived at Dalat, another city surrounded by steep mountains, a city heavily French influenced, which you can see in the architecture of their older buildings. A truly beautiful place and the location of the Republic of Vietnam's Military Academy. This is their equivalent of West Point. Here there is a lake landing pad which we used on a consistent basis. This is where Bruce, my roommate, almost ended up swimming after he tried to take off with a heavily loaded bird when his engine RPM started to bleed off (hovering over water requires more power due to reduced rotor downwash, less cushioning effect) over water. Fortunately, Bruce managed to just barely hover back to the pad. Interestingly, we would often watch the locals wash their elephants here. Most of these animals were trained for logging and, like most hard-working animals, you could see they enjoyed their cooling bath opportunity.

Amusing, but that two-hour flight over these impenetrable jungle mountains that we were so concerned about from Ban Me Thuot to Dalat would now be repeated as we now took off from Dalat to our sure return to Phan Thiet. More mountains, additional jungle, but it was in our own area of operations. We felt the home field advantage, and mental reassurance of the familiar. I would fly this route numerous times as the year progressed but would always dread that Ban Me Thuot run. That lone thought of never being found is like extra baggage imaginatively there for the ride. I cannot speak for the other pilots on how they felt, just for myself. Home is where you lay your head at night, even if it's Phan Thiet/LZ Betty. Just a typical day starting at about 5am and ending just after 8PM...more to follow.

Mountain Terrain

6. CHARLIE CHARLIE

Now having been incorporated into the first Flight Platoon for over a four-month span of time, I was no longer considered a new guy, but just one of the everyday co-pilots who would receive their assignment from the nightly mission board. Every night the Platoon commander would receive the missions for the next day from Operations and then proceed to his room and grease pencil his choices of the pilot pairings assigned to these missions. Immediately after these selections, the platoon commander would post the mission board outside his hooch door for the crews' first glimpse of their next flying day. We always had dedicated missions like the bus runs, resupply and "Command and Control" (CHARLIE CHARLIE).

This last category usually had an assigned PC to ensure daily continuity and familiarization of the unit being supported. The assigned PC would fly with the unit Brigade Commander to all the locations of his field units, and by having the same PC, the daily orientation was minimal. The CC ship was set up for the ground commander with a special console secured to the cargo floor which contained extra radio communications, including SIPER (encoded frequencies). The ground commander would often fly with his Battalion and Company commanders to get the aerial overhead embodiment of their unit's position and defenses.

Our CC assigned Pilot in Command was Jimmy Lynch, another New Yorker and die-hard Mets fan. On October 16, 1969, the NY Mets had just won the World Series in game five, some of the NY pilots were celebrating this world news with extended nighttime drinking. Now being a Yankee fan and not really a true sports enthusiast, I was sleeping when Jimmy and the New York gang came bursting into our hooch to celebrate this triumph. It would be a very long day tomorrow, but Jimmy and I would both be there for I was assigned as the copilot for Jimmy Lynch. So, we never really got mad at each other, we were all brothers stuck in the same dream, in the same location, with the same goal of eventually getting home.

For some unknown reason, I was assigned to fly with Jim as his co-pilot on Charlie Charlie for what seemed to me a very long period to fly on these Charlie Charlie missions. No complaints, every day you flew was a learning lesson. With this mission, I was fortunate to engage in a large workload with numerous landings and takeoffs in rugged locations, coupled with the observations of a ground commanders' operational decisions. Radio transmissions were our connection to the ground units, and you would be changing frequencies constantly to a new push (term used to change to another frequency); so many different frequencies while constantly using your authentication code book to contact the next unit. FM Radio frequencies may not be secured and often could be monitored by the enemy. If a landing was to be initiated, we would request that they pop a smoke grenade to confirm their location and have visual verifications of the landing zone. The grenades came in four colors: green, yellow, red, and purple and after they were thrown and spotted by us, we would ask "confirm color" they would respond with the color they threw and we would say "Roger," authenticating their color.

On this day, we would head south to a large mountain in the Ham Than District with the Brigade Commander, call sign BLACKHAWK 06. He was right out of the movies, shaved head and go go go. We started to check out some of his units' areas with overhead flights and occasional landings so he could add the personal touch to his troops. I think his unit was mechanized Infantry using APCs (armored personnel carriers) to traverse the lowland brush forest. As we now scouted to the south on the edge of what I was told was a sacred mountain, we spotted several individuals on the north side of the mountain in a small clearing close to the coast. This area was a free fire zone, which meant you could treat any unknown personnel as the enemy, now snared in this death zone, too bad for them.

At that time, the gun cruiser USS St Paul was cruising just off the coast near this location and was available for artillery support. So many units, so many frequencies, including the Navy, and believe it, we had all those frequencies on a daily basis. BLACKHAWK 06 called for a fire mission to the St Paul. This is accomplished with a SALUTE report (size, activity, location, unit, time and enemy). The ship was given the eight-digit grid coordinates of the enemy location and began their gun solution computations to fire on the suspected enemy. Our helicopter took up a racetrack pattern that was parallel to the gun target line from the ship to the enemy. This flight maneuver was imperative to keep us out of the planned line of fire that was about to commence while giving us a unique observation of the pending action.

The St Paul gave us a heads up that fire would commence shortly, then announced, "shot out". That combat verbiage let us know that they had just fired a three-round salvo (bombardment) from their eight-inch gun turrets. They had launched three rounds, weighing two hundred pounds each, containing high explosives encased in their steel jackets. The St Paul next transmitted a warning of "splash out", this universal artillery verbiage told us the rounds were about to impact.

We watched intently when all three rounds impacted the target area with unbelievable force, stunning as that familiar pulse of the compressed air wave proceeds the plume of pulverized earth, mixed with the black tint of rapidly burned explosives.

Those suspected insurgents were gone for sure, they probably, in that last fraction of a second, never heard the whistle of those immense rounds just before impact. BLACKHAWK 06 called to the St. Paul to "Fire for effect", as if the first three rounds didn't have any effect and there may be enemy remnants in the nearby tree line. This was a common practice because, unlike in today's advanced artillery shells, the artillery fire of this period was not always as accurate, as I will soon explain.

The St Paul announced again, "shot out" as we again waited for the next three monster shells to impact. The St. Paul announced, "splash out," we acknowledged with "Roger." Shock! Those three rounds landed to the right of us, through our racetrack pattern, so far off course we instantly called for, "cease fire, cease fire." Had they impacted our ship there would be no pain, like raw hamburger meat hit by a high-speed golf ball.

The St. Paul quickly responded, acknowledging that they had an error with their ships fire control computer, which we immediately confirmed, and called for an end of the fire mission.

As we departed the area to continue our CC mission, our conversation was centered on this event with laughter about the Navy, and how at least the first three rounds took out the enemy. My God, life was so cheap.

We landed after another late day, knowing more CC missions were to follow. There would be one other CC mission, very sad and forever imprinted in my memory.

CHECK RIDE

Picture from USS St Paul depicting one artillery round from their triple gun turret, imagine three of these in flight. (photo copied from St Paul web page)

7. Ambush

Flying CC (Command & Control) ship with Jim Lynch as my PC was always an early start, late finishes, and lots of neck pain, literally. Flying Command and Control requires many hours of left-hand circles while the Brigade Commander Blackhawk 06 continuously looks down and critiques his units from his aerial observations. As the co-pilot, I sat in a fixed armored seat on the right side of our aircraft for hours upon hours, only seeing high sky. That takes its toll, with headaches and stiff necks while I prayed for a fuel break, or anything to fly straight and level. Maybe that was why other co-pilots politicized the Platoon Commander to avoid this daily mission. I should have been more social, I'm a slow learner.

This date we were flying north of Titty mountain, directly over Highway One. That road is the main coastal highway which runs north and south through the entire country, all the way to Hanoi in the north. I would fly over this highway almost every day I was in Vietnam. The road is a museum of war with old Japanese tanks and French tanks sitting on the side like a ghost signpost saying, "beware of this road." Ambushes were the norm there and many soldiers have been killed or injured. Unfortunately, it was still the main supply route for our military to support the numerous bases like LZ Betty. One of Blackhawk 06's missions as

Brigade Commander was to have his troops protect and patrol this road on a daily basis. As convoys of supplies moved, they were escorted by APC (armored personnel carriers) and in some cases tanks. Vietnam is a country that has been in some kind of conflict its entire history and this road has seen its share of conquerors and the defeated, we were just the current contestants.

We were on station ten miles north of Titty Mountain, with Highway One directly below as our left turn orbits began. Beneath was a convoy moving slowly northward with troops and supplies for someone. The first indication something was happening was that the radio chatter quickly became panicky with stress, which was immediately injected into the ground radio operator's voice. The convoy was under attack directly below us, in daylight. How long have they been waiting, hiding for this opportunity, intensely focused on causing concentrated damage to their foe.

Attacking a road from the dense forest brush on each side was common, effective, and the surprise of the attack accelerated in triggering confusion. With enemy rounds coming from all directions the convoy commander had to first coordinate a defense, then if possible follow his chosen course of action by an immediate counter attack plan. When being engaged by a well-prepared enemy ambush, you don't immediately know the size of the enemy force, and in a convoy the commander's location could possibly be at the front or rear, now trying to relay his commands over a burdened FM radio system from stressed out unit radio operators.

Blackhawk 06, the Brigade Commander sitting in back, jumped in on the frequency. Wearing his headset and microphone, he was requesting updates. You could now see the intensity of the firefight with green enemy tracers and red friendly tracers ricocheting into the sky. Our troops have very powerful weapons, including the M2, 50 caliber machine gun with a large caliber round cutting deeply into the low forest. These rounds have so much inertia they will sometimes strike the earth and ricochet skyward, still burning their tracer brightly red a thousand feet into the sky after the initial ground contact. The roadway below turned to dust as vehicles and gunfire continued to erupt in every direction.

We were an airborne audience flying in a fifteen-hundred-foot pattern to the tragedy below, soldiers had been wounded, seriously wounded.

The convoy called for help, and back at LZ Betty, the Tiger Sharks of the 192nd gunships were scrambled and would be on station within fifteen to twenty minutes, but by then the enemy would have dissolved deep into the low forest brush. They would be too shy when the gunships arrived, knowing that their presence can be observed quickly from the sky.

The convoy commander updated Blackhawk 06 of his wounded. Some were so serious they needed immediate evacuations. Medevac was called. Medevac are UH-1H helicopters with a specially trained crew, which includes medics, to start treatment the second the wounded are placed onboard. These special aircraft are the direct descendants of those early trials of helicopters during the Korean War and are the prevailing reason why only 58,000 soldiers died during the Vietnam conflict, instead of maybe 100,000. Many consider them Angels from the sky, and I would agree. Their helicopters are unarmed and the enemy sees them as easy targets, no rules, no mercy.

Blackhawk 06 believed Medevac would be too slow for his boys and requested that we land on the highway to try and get these soldiers/boy's, someone's son, father, and brother, to the aid station at LZ Betty in the shortest amount of time. Now Blackhawk 06's rank is a full bird colonel, next step-up would be brigadier general, but this decision was the PCs, as commander of this machine, Warrant Officer Jimmy Lynch, who is in charge of our ship. It was a no brainer, we GO! Our approach was to the front of the convoy, the ground troops had already deployed smoke to help us judge the wind, always a factor if properly used, a tool/edge when heavily load aircraft need to take off.

We landed in a puff of dust and this would take some time. Now sitting there on the highway, a prime target, but it was the job we accepted, and probably one of few times we truly feel needed. Convoy soldiers arrived with their wounded, one soldier could climb onboard by himself, he was bleeding from a badly mangled hand, the blood had exceeded his bandage and he was strapped into a seat by the crew chief, enduring the pain. Next, a soldier was carried and placed on the cargo floor, bleeding from his neck, conscious, he was shirtless, which was probably removed by his comrades to apply first aid. The final soldier was almost thrown into the nylon seat, urgency reflected from the soldiers carrying their comrade.

This seat which had been set up ran the width of the helicopter. Blackhawk 06 sat next to this young soldier, talking to him with his hand on his shirtless chest, this soldier's skin color was like a light wax. There was a wound to the left of his breastbone (sternum), the size was shaped like that of a fingernail on your pinky finger, a deep wound. This young man was dying. As I looked back, you could see he was gone or going. Nothing further could be done as Blackhawk 06 touched his chest to comfort and give solace to this hero. A connection from a commander at his zenith to his soldier at the far end of the US Army rank structure, a young sergeant or corporal, a boy under his command mortally wounded as our helicopter took off toward LZ Betty. That soldier's Family's life had changed, they would not know for maybe twenty-four hours before this pain reaches them that this young man was gone.

We were airborne, flying at maximum speed, gauges to the max VNE (speed never to exceed). Jim was on the radio alerting the control tower that we were inbound with seriously wounded, clear the way. The UH-1 has a very large windscreen, made of very thick Plexiglas, a fixture we as pilot's love. The more you can see, the more reference you have, greatly appreciated, "Thank you, Mr. Bell Helicopter." I stared straight ahead, watching our course when I noticed something on the inside of the wind screen. I was wearing nomex aviator gloves which have a pig leather underside designed for a more sensitive feel, which I now used to take my left index finger and gently wipe the inside of this glass, surprised by what I now discovered. These wounded soldiers' blood had now atomized because of our open doors circulating airflow through the ship, causing a thin film onto the inside of our wind screen. I could draw a line in the human blood residue, blood of these unfortunate victims of war, one whose future was forever ended on that ghost highway with all those ancient warriors.

Landing at the medical pad at Phan Thiet (LZ Betty), the pad was surrounded by medical attendants and doctors geared up to give these soldiers their best chance of surviving. Our job was not done, we were off to hot refuel, then back to pick up Blackhawk 06 from the medical pad to continue Command and Control.

It was too easy remembering this mission, and I often will say a prayer for these soldiers and their families, mostly when I would be driving alone early in those morning hours when the sky is painted with God's own palette, those quite times you can be thankful for what you have, knowing what others have sacrificed.

8. LRRP

Now a seasoned co-pilot, having participated in every varied mission on a daily basis, you can now distinguish which are the most dangerous and challenging missions.

One of the missions the 192nd Assault Helicopter Company was dedicated to was supporting Long Range Recon Patrols (LRRP). This type of mission involves transporting an elite team of Army Rangers who had been assigned to the 75th Ranger Regiment into the mountain jungles for reconnaissance, ambush, intelligence, and a whole host of some serious warfare assignments. They were usually four to six-man teams, armed with every conceivable weapon which can be man transported. Their rucksacks were ponderous, like the chains of Marley's Ghost in a Christmas Carol. They would be self-sufficient for several days while roaming jungles and mountains so rugged and dense you wondered if they hadn't been digested by the vegetation.

These men would be loaded onto a slick at LZ Betty, all bent over almost ninety degrees to facilitate carrying these huge rucks loaded with ammo, food, socks, rain gear, radio's, extra weapons, explosives, grenades, smoke grenades, flashlights, and whatever personal links to home and family which supports them through the mission. True warriors, fearless, camouflaged to the max, painted to blend in with an unconventional head gear in place of helmet, anything and everything to be part of the jungle.

I was still twenty-one and, thanks to the Army, I had been exposed to so many new experiences which were totally unexpected. For this primary reason, I always recommend to young adults, if you're lost in life or cannot determine what path you should take, check out the military service. They will expose you to so much and open your eyes to this immense world.

Having joined the army in May of 1968. I travelled eventually to Alabama where I was thrown into an anachronism, a condition I had only heard about. One day while stopping at a small gas station, I noticed a sign for the rest rooms which stated colored. I started to pay attention after that and again noticed at the local movie theater they also had a separate entrance for colored. I was from the northeast and I never imagined such a thing existed, that's how protected I was. Not berating the South, they were some of the kindest and sweetest people I would ever encounter on my journeys.

My mother and father never had a bad word about black people, which we now called Afro-Americans. The reason I mention this at all is because I would learn my greatest humanity lesson while observing these Rangers, unburdened by race, ethnicity, or any discriminating characteristics to separate them into some category. These men, combined of all races and all true Americans, would forever reinforce my definition of our America. I realized, although of Irish and Scottish heritage, I was an American first. From this time on, if you were an American, I would consider you a brother or a sister evermore.

LRRP missions involved five aircraft. The first was a Cessna fixed winged L-19/O-1 Bird Dog, which was a liaison and observation aircraft (FAC, Forward Air Controller). This aircraft could adjust artillery fire as well as perform liaison duties such as serve as a communication link for LRRP soldiers, and often carried four rockets on rails under the wing to mark targets for fast movers (jets) dropping bombs. This lone pilot would loiter over the insertion area while assisting the helicopter with this heroic team onboard to maneuver to the exact grid and LZ for the drop. Bird Dog aircraft would also be the first ones to receive any alarm call if this small team was in peril, an event that happened more than you might expect. What aided this small fixed wing aircraft was its ability to stay on station/location for up to five hours of overhead coverage.

The next assigned aircraft were two UH-1C gun ships from the 192nd gun company, call sign TIGER SHARKS. These were modified UH-1B or UH1-C models, smaller than the troop carriers, but heavily armed with two Gatling machine guns capable of firing five thousand rounds a minute, side mounted 2.27 inch rocket launchers with seven rockets on each side capable of delivering a five to seven pound explosive warhead, and inside the small cargo cabin two M-60 crew mounted machine guns slung by bungee cords for maximum deflections and defense as these birds would pass over an enemy location exposed to fire from their rear or side. On LRRP missions they would assume a standoff position to be within striking distance if immediately needed, a compelling testament to the hazards of these insertions.

The final assemblage of this mission would be two UH-1H model slicks, larger than the gunships and with a stronger, slightly more robust, engine, call sign POLECAT and referred to as slicks because they were only armed with two M60 machine guns. These helicopters primary usage was transport, to include this precious cargo of human excellence. Only one ship carried the LRRP team, the other was the rescue bird in case the mission went horribly wrong. This rescue ship would also maintain a standoff distance far enough to prevent the enemy from detecting an operation. The final loaded ship, being a UH-1H and having one of the most distinguishable sounds in aviation history since Orville and Wilbur stirred the sands at Kitty Hawk, would follow their map, carefully reading the contour lines and grid coordinates to a pre-planned drop zone while the O-1 Bird Dog was high overhead in constant communication and would assist with updated countdown of the distance to the landing site (two Clicks, one Click, a click being 1000 meters). Sometimes you would feign an approach to a separate location to possibly confuse the enemy with the aircrafts distinctive low pounding wop wop sound which would reverberate in the mountain valleys. This was the enemy's turf and you wouldn't want to land on a bee's nest of scared NVA or VC.

The jungle from a helicopter is a wondrous sight, towering trees, rock cliffs covered in lush green vegetation, waterfalls you only have seen in resort photos, small areas with high grass, bamboo seventy-five feet high, monkeys jumping from tree top to tree top, and the occasional larger dark grey stones in a barely open patch that you soon realize are wild elephants. There will be times you just can't get the machine low enough for our troopers to step off, so they must jump.

On one occasion, the drop area was large enough to hover down to what appeared to be a very low level of a vegetated green field, but the LRRP team would have to jump those last few feet. I will never forget this moment of courage, this moment of mental devotion few could complete or even comprehend it. Looking so close so easy a three-foot leap onto some high grass of the jungle floor what happened next was carved into my memory of the meaning of **"RANGER"**. The team leader jumps and doesn't stop. He actually punches a hole through this magic green carpet maybe ten to fifteen feet totally unexpected, it had to hurt, hell, it hurt me to watch. An overburdened soldier carrying maybe an extra eighty pounds like a lead sinker into the ocean straight down to the bottom.

His team watching and without any hesitation now jumps one after another until we were empty, they are now quickly devoured by this living biomass, off onto their pursuit of CHARLIE. Army Aviators have a singular utterance for such a display of warfare courage, **"BALLS"**!

We were out of there, now climbing with more power than we needed after the weight loss from the team's insertion. We were headed to home base and the next assignment. That picture will forever be imprinted in my mind. American soldiers at their finest, people should know of such acts. This type of scenario would be carried out hundreds of times during my assignment at LZ Betty, and one we learned to enjoy for the challenge and pride as Army Pilots.

I wrote this as one small flavor of what these men did, later as a PC myself, I would be responsible for these men. As I have previously mentioned, my co-pilot training and learning process was crucial and I thank those that came before me for their unselfish attention to my internship.

BIRD DOG OV-1

RANGER

9. TAGI

Late November or early December 1969 was the dry season for our area. I have been scheduled to fly as a copilot with a new Pilot in Command, Tagi. I would become a pilot in command by the end of December or early January. On this date's mission, I was to assist the new PIC (pilot in command) who would normally be assigned to the MACV bus runs of ferrying multiple passengers from village to village, a fairly safe flight and training exercise to instill confidence, coupled with command presence and the decision making processes needed to confront challenging future flight assignments.

Lee E. Tagi graduated from WOC (Warrant Officer Class) Class 69-1 and had arrived maybe a month or two ahead of me in Vietnam. Tagi was assigned to the first flight platoon and his roommate was Larry Hupe, Larry was the officer who I first met on my arrival to LZ Betty. Tagi was an amusing guy, loved smoking, drinking, and laughing, not unusual behavior for most warrant officers, probably exaggerated by the ten plus months of constant harassment/bullshit we all received during flight school training. A type of catch up entitlement we all believed was acceptable and proper, similar to Mom and Dad going away for a weekend and telling you to behave.

Tagi was always in the forefront of making us laugh, both he and

Hupe early on desired to transfer into the gun platoon before their time was completed in Vietnam, for reasons unknown to me except for more down time. Flying gunships restricted you to always being based at Phan Thiet, no real long range or overnight missions, many days of not flying. I preferred some variety with the opportunity to fly as often as possible, but this pace would take its toll slowly/ furtively wearing both the body and spirit no matter where you were assigned.

I cannot remember all the details of this mission but will relate the one leg that was unpredictable and unforeseen. Our first pickup stop was inside the City of Phan Thiet, minutes from LZ Betty. It was a small pad or large courtyard in the center of the city with some type of military presence for security, surrounded by stucco buildings. Upon landing in these populated areas, they always had a very unique odor which I have always remembered as rotting vegetation blended with charcoal burning fuels, combined with whiffs of human waste due to unsanitary conditions, and all finely tuned with the heat and humid conditions of Vietnam, now forever stored as one of my most unusual memories, not déjà vu, but an experience etched stone deep in my brain.

Completing the first scheduled stop, we now began a sequence of flying to a chain of locations and maybe the next would be the very small village of Ham Tiem, just northeast of Phan Thiet on the coast, possibly a future candidate as sister city to Carmel, California. It had that type of potential, the same routine military/ civilian's pax (passengers) both male and female who had been preapproved. Attired in their traditional coolie hats and silk shirts and pants, sometimes in western dress, many of them heavily perfumed or cologned, often coupled with their bags of all shapes, some cargo or mail which was loaded onboard. The MACV bus routes were by far the safest and fastest form of transport in our particular area of operations, but these people were our passengers.

Flying northbound while climbing to our safe altitude of fifteen hundred feet, again the altitude determined to avoid directed small arms fire, absolutely the standard. Light clouds with maximum visibility as we traversed the low jungle mass below the Lee Hong Phong forest, a coastal plain, geographically located between the South China Sea beaches to the east and Highway One to the west. This forest was the perfect hideout for VC/ NVA soldiers with its many low trees and wilds abundantly grown into an entangled mass of knotted vegetation so difficult to penetrate it received special attention from U.S. forces. So formidable was this green badlands that the entire area had been sprayed by the US Air Force with the defoliant Agent Orange, a powerful chemical defoliant which would unfortunately greatly increase the casualty rate of this war long after our country's departure.

Apparently even this application was insufficient to move our adversaries from their sanctuary, so sparing no expense, the US Army decided to carve wide swaths through this knotted forest with Rome Plows to allow easier access for our APCs and tanks to discourage our constant and determined antagonists. Similar to the Roman assault on Masada—my analogy since the name Rome brings to mind such ancient warriors" where they were so determined to build a roadway up a mountain to continue their siege. It is easy to associate these lessons handed down through history.

Land clearing was another important job of the engineers in their operational support role; in fact, engineer methods of land clearing gained wide acceptance as among the most effective tactical innovations of the war. As techniques evolved for the employment of land-clearing units, these units more and more became the key elements in successful operations aimed at penetrating enemy strongholds, exposing main infiltration routes, denying areas of sanctuary, and opening major transportation routes to both military and civilian traffic. Engineer land-clearing troops on many occasions formed the vanguard of assault forces attacking heavily fortified enemy positions, while even under ordinary circumstances their use in clearing the jungle ahead of tactical security elements placed them routinely in a position of direct vulnerability to enemy action. These engineer troops rapidly developed a zestful pride in the importance, difficulty, and hazards of their occupation, and while their deportment and appearance sometimes fell short of normal standards, their spirit, courage, and persistence under the most adverse conditions entitled them to a substantial claim to elite status.

In the vocabulary of U.S. forces in Vietnam, Rome Plow came to be synonymous with land clearing. Of all the various types of land-clearing equipment tested in Vietnam, the military standard D7E tractor, equipped with a heavy-duty protective cab and a special tree-cutting blade manufactured by the Rome Company of Rome, Georgia, proved to be by far the most versatile and effective. The tractor took its name from its most imposing feature-the huge blade on the front.

Le Hong Phong Forest a Difficult Operations Area

Rome Plows Clearing Forest

Rome Plows in Action

Agent Orange
/noun/

1. a powerful herbicide and defoliant which contained dioxin, a toxic impurity which caused serious health problems, including cancer and genetic damage. It was named after the orange stripe on its container.

Agent Orange Herbicide Barrels

Helicopter with Agent Orange Spray Rig

CHECK RIDE

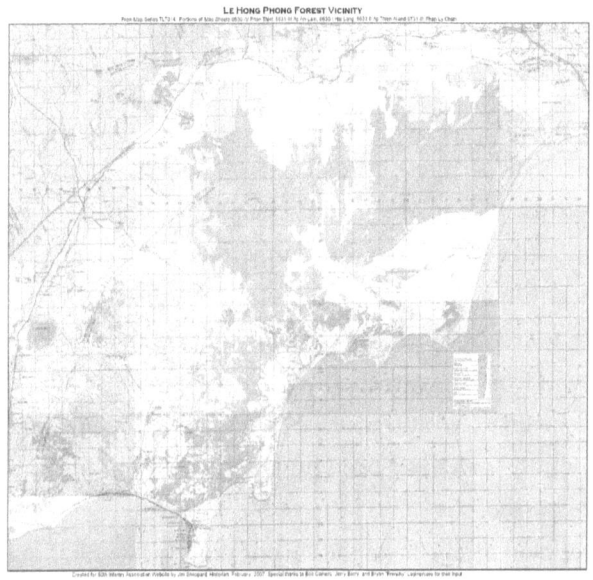

Grid Map Le Hong Phong Forest

It was one of those tranquil days, good company, a defined mission which was familiar to all four of the crew, a slow grounder to the shortstop. Typically, you could be flying an eight-hour day plus, not including one hour for flight prep and one hour for post flight duties, especially if you went north to Nha Trang. Tagi was a good pilot, a true warrior who never withheld his desire of going to the gun platoon. It was a pleasure to share this day with him as I listened to his dreams. Such verbal exchanges often reminded me of my friend John Wright who we lost in October on a night gun mission.

There was one incident where Tagi was the central character in a medical scenario not expected, which effected every member of the 192nd Aviation Company after he had transferred to the gun platoon. We all noticed while Tagi was holding a beer in unison with the always present cigarette that his eyes were yellowish. You didn't have to be a medical professional to know when a child was sick and clearly something was amiss with Tagi. Encouraged to visit the medic, news quickly spread Tagi had hepatitis and due to the contagiousness of this disease, the entire company had to get a gamma globulin injection, a very thick serum, given with a large gage syringe to enhance the immune system, administered directly to the gluteus maximus, dreadfully painful. Tagi was our patient zero and focus for numerous comical comments till his end of days at Phan Thiet.

Free fire zone was a term used during the Vietnam War. It was a designated area in which any weapons or weapon systems could fire without having to coordinate with their headquarters. A free fire zone is an area where all friendly forces had been supposedly cleared, and any remaining people were hostile.

Interpretation of free fire zones in our unit was word of mouth passed down from Pilot in Command to Copilots since I arrived in country. No formal explanation of a defined procedure and no FM 6-20 (Field Manual) was ever seen by me or my fellow pilots that I know. Common knowledge was that the whole of the Le Hong Phong Forest was a free fire zone and the honed skill of reading our grid maps correctly was required.

Now on mission, flying north northwest over the Le Hong Phong our passengers were nestled in their respective seats, enjoying their unobstructed aerial view of the bramble forest below, never expecting what was soon to transpire. I cannot recall which crew member first observed the subjects in the forest below but flying overhead your instincts are similar to a hawk or other bird of prey, your eyes continually scan the environment, occasionally detecting the slightest movement.

Alerted, we immediately went into left circular orbit like vultures to examine our prey beneath. Gliding overhead in this death orbit, we were safely protected by altitude but advantaged with our vertical scrutiny. Clearly three subjects were located in the woods, all wearing black silk pajamas, not necessarily combat clothing since this would be their version of Americans wearing jeans. Two had straw coolie hats, common/typical for the hot sun-drenched days in Asia. Tagi quickly reviewed his grid map as I maintained his requested orbit, the three subjects were immediately aware of our ship, looking up. Friend or foe they knew they have been observed and I can still only imagine their thoughts, but I would suspect they were leaning to the negative side of possibilities, army helicopters had a threatening repute.

It appeared to me those men/boys were gathering wood, most possibly to make charcoal. There was no doubt our location was outside any friendly forces but only maybe two hundred yards inside the free fire zone, so, such confirmation was required to ensure and prevent fratricide. Tagi was now on the FM radio (fox mike) checking in with both a flight following frequency and a DivArty (Division artillery channel) frequency, both would or should have current locations of possible friendly units in this area. As we continued our orbit, the characters of our attention below began to move westbound on a dried-up trail, knowing our orbits overhead had completely focused in on them.

Tagi/future gun pilot/PC made a command decision after receiving radio authentication confirming that we were clear of friendly units. The three subjects were unequivocally in a free fire zone and we would engage them from altitude. Engage, a soft word for attack and kill. The three unfortunate souls below whose intent would never be determined, especially from fifteen hundred feet were now targets. I was not totally onboard with this decision, but Tagi was my PC and my job was to be part of our team. He was not outside of our ROE (rules of engagement). I still always thought these subjects were just gathering wood and their location was unfortunate, but war is governed by location, consideration, deliberation, contemplation followed by reflection on such actions which will follow for hundreds of years of regurgitated interpretation of history.

We continued in a left-hand orbit with Tagi now at the controls as he ordered his crew chief on the left side to "engage" the enemy with his M60 machine gun. A very capable weapon whose rate of fire is 500 to 600 rounds per minute with every fifth round a tracer making the rounds visual path extremely observable. Sitting in the right pilot's seat I strained my neck to the left waiting for the first burst from the chiefs' machine gun, knowing these three subjects would be drilled into the earth with no explanation of their circumstance just their brutal demise in the woods.

As the chief placed pressure on the trigger, that now so familiar pulse of exploding gun powder ran through the ship as I instantly saw the muzzle flash while the first rounds arched toward their prey. Maybe three tracers guided my eyes, with their burning red tails, in-between four unseen copper clad steel projectiles, each round travelling at 2000fps (feet per second). I could see the first rounds scatter around our targets with one tracer astonishingly going directly between one of the subjects' steps, kicking up some dust, mind-blowing. The three didn't panic but kept a steady pace treading toward that invisible line of safety where the free fire zone ended. I was amazed at their composure as Tagi continued his orbit. The chief's gun had jammed, not unusual if not cleaned properly or just parts worn out. The chief made his second attempt with that one pop as the gun jammed again. Sometimes the Gods watch out for the other guys as our prey entered the grey zone of good or bad decisions'.

Tagi realized the situation had changed as we both quickly discussed the dilemma, followed with the righteous conclusion to disengage as we now aimed our aircraft back on course to conduct our humdrum MACV bus run. Laughter, jokes, and the humor of youth scrubbed the mind of the seriousness of this action as we continued the mission. I hope the gift of a potentially long and fulfilled life given to the three unknown targets was cherished as I remember their unplanned fortune.

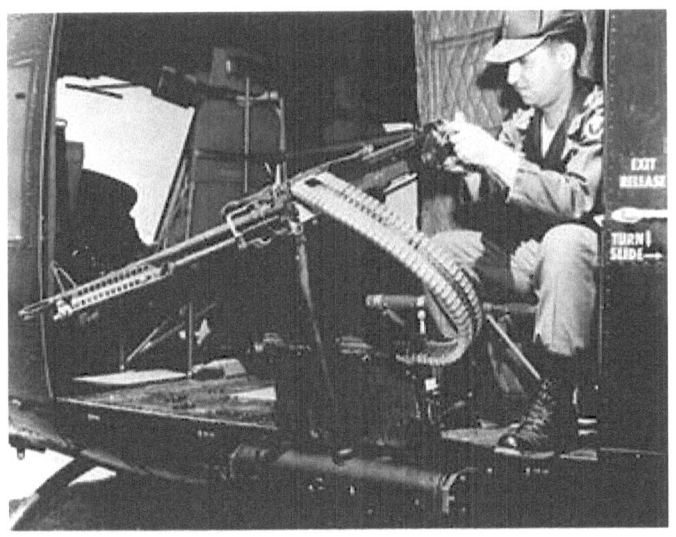

M60 Machine Gun Crew Chief side

10. The First Time

I cannot remember the exact date and time, but I do remember it was early on during my apprenticeship as a co-pilot, possibly late July of 1969. I was flying this day with a new pilot in command, Mike Porter, a young small man with a great attitude, friendly, talkative but informative. Learning, always learning. It was late morning, maybe early afternoon, I think we had already completed some resupply/log (logistic) missions when we returned to LZ Betty, only to be now reassigned a most unusual mission.

Normally, the Company's missions ranged from bus runs, resupply/log, admin, LRRP insertions, combat assaults, and the occasional WTF. This was closest to the last category when there was only one passenger, a lean built man, maybe 5' 9", completely dressed in black silk pajamas associated with the everyday garb of a common Vietnamese field worker. This ensemble was garnished with the iconic coolie straw hat and a pair of Ho Chi Min sandals cut from old automobile tires.

Our passenger was sporting a pale green sub machine gun with the ammo clip on the side, and one last splash of impressive embellishment...our man had a very large knife, that's it. So, if you want to believe in Rambo, yes "Virginia" he does exist.

Mike briefed the crew on the bizarre mission for our mystery guest.

I was told this guy is a Navy SEAL (Sea Air Land) and our mission would be a single ship insertion of him about ten miles down the coast. Having only been in the Army over a year plus, I had heard of SEALS, but at that time in my early soldier orientation, I never really knew what they were about. Looking back on this now, with my knowledge of this elite unit gleaned over my past forty plus years in the military, it's easy to accept their deserved future reputation for the unorthodox. But unquestionably, they had a low and limited profile presence in the Vietnam war in my operations area.

Phan Thiet was a prominent city on the east coast of Vietnam and LZ Betty was located just south of the Cay Ty river entrance, on a raised plateau sitting on red sand cliffs overlooking the South China Sea. Just south of our base was a low brush forest and the usual location of directed enemy mortar fire which we received on a regular basis. Also, south along the coast was a large sand dune area that would, with a bit of imagination, make you think of the Sahara. Having revisited this area with the internet, using Google maps, these dunes are now quite an attraction and Phan Thiet has turned into one of Vietnam's world-renowned resorts areas. Who knew?

We departed LZ Betty with this insane passenger, a lone man going into Indian country with barely anything to ensure his survival. My best assumption was that this courageous soldier would scout the coast for the heavy gun cruiser St. Paul, a former WWII ship with eight-inch guns capable of reaching targets many miles inland from the coast. Our flight was south down the beach, low level, when this fearless cat directed us to turn right up a small brush covered ravine where we could locate a place to drop him.

We quickly landed in a one ship patch of low shrub, and out he went, off into the low dense mangle of a beach forest, like a released animal back to the wild, never looking back, escaping his cage to quickly melt into the bush. We then proceeded back, flying north, staying low crossing the dunes area when the crew chief yelled, "We are taking fire!" Mike acknowledged that and maneuvered quickly as the gunners raked the brush near the dunes.

Frankly, I didn't hear or see anything, but hey, I was new here and should learn. Stealth is not associated with helicopters, especially the UH-1; they gave us helmets for two reasons...one crash protection, two hearing protection. I'm sure someone had fired at us on many other occasions that we will never know or need to know.

As for the pax (passenger) that we had just inserted, I never heard anything else about him, I presume he may have been picked up on the coast one night by Navy boats and always wondered how we find such soldiers, but I would go on to see many more young fighters like that and always will be in awe of their feats. Time was moving slowly; adventures would be replaced by unembellished reality.

Ho Chi Minh Sandal

11. Flare Incident

Gradually acquiring much needed pilot skills and on this night after a considerably extended flight day, our ship was assigned primary flare aircraft. There's no such rule as "crew rest", a term invented in future years of Army Aviation. My PC was Mark Clements, a gruff no BS attitude guy who gets the job done, a true Pilot in Command and he would instruct me competently in the proficiencies I would need to resort to and depend on after such men left. I loved to fly with Mark; he would let you handle the controls and never be shy about telling you what and how you should be doing maneuvers.

I believe it was Mark who told me one day while we were flying over some rice paddies at fifteen hundred feet cruising along in the daytime somewhere when he ordered (of course mixed with expletives) me to take my Smith & Wesson .38 caliber revolver from my shoulder holster and shoot it out the co-pilot's window, thoroughly instructing me to be looking straight down the barrel of the weapon. I snapped my head nimbly, now conspicuously glaring directly at Mark and articulated what any normal person might reply, "Are you serious?" He replied, "Yes, just fucken do it." Following Mark's instructions, I proceeded to fire my first non-range round in Vietnam, firing one shot while staring directly down the barrel of my issued sidearm as directed by Mark.

Watching intently, you could actually see the bullet leaving the barrel and arching to the right as we sped along. A most impressive lab experiment with the instant realization of **inadequacy**. The crew, hearing the pop, called over the intercom system, voices piqued, asking, "Are we taking fire?" Mark cued them in on my field combat education lesson. Mark then illuminated me that there were lots of weapons for sale throughout the Aviation Company, finishing with his emphatic, **"Get one!"** The next day I purchased a carbine with four twenty round clips and later also purchased a jungle shotgun with a bandoleer of shells. Believe It, I felt more secure, knowing I would stand a fighting chance if we went down.

Every pilot discerned that this issued weapon/revolver was mainly issued for the purpose of self-inflicted death rather than face the possible capture and torture by our enemy. I hate to use this analogy/cliché, but this is the one-time size does matter, sorry ladies, but so true.

Charlie hated helo crews, and if you went down and survived whatever caused the event, chances were good that you would be rescued immediately by our troops. However, time was always working against you and if the rescue was delayed, your chances of living were zero if captured, better to die in the wreck. In the South, Charlie rarely took prisoners.

Flare ship is a UH-1H configured with cargo doors pinned back and specially rigged for night flare drop operations. The actual setting up of the aircraft took fifteen minutes and it remained in the arranged standby rigging until daylight arrives.

Strapped to the cargo floor were two large aluminum open topped buckets which sat just outside the cargo compartment of the ship, secured by nylon straps. These buckets carried a number of MK24 magnesium flares on each side of the aircraft. The flares had a timing fuse and wire lanyard to activate them. They also contained a small explosive charge. As they were tossed by the crew chief and gunner into the blackness of the Vietnam night, the charge activated the process to eject the flare from its aluminum container. The flares were just over three feet in length and about eight inches in width, containing a medium sized parachute to slow its descent as the magnesium flare was ignited, which then gave maximum illumination to our gunships below. Every evening a flare ship was prepared just in case the guns were called out for units in contact. I would participate in this type of mission several times during the next year, one future one would almost be fatal.

The alert horn sounded for the gunships to immediately launch in support of our troops in contact. As primary flare ship, that included us. We grabbed our essential flight gear and ran blindly into the dark to launch. We were quickly airborne, climbing to five or six thousand feet, depending on where we are needed. Navigation in Vietnam is a challenge, we had no VOR (**VHF Omni Directional Radio Range (VOR)** is a type of short-range radio navigation system for aircraft, enabling aircraft with a receiving unit to determine their position and stay on course by receiving radio signals transmitted by a network of fixed ground radio beacons), only one ADF (A **non-directional (radio) beacon (NDB)** is a radio transmitter at a known location, used as an aviation or marine navigational aid) but they only work if you have a good signal.

The cabin doors were open because of the flare buckets and it was cold, for every thousand feet you climb, the temperature drops two degrees Celsius. Quick math, if ground temp was 68F, our temp was now around 48F. With a wind chill added to the temperature, it was very cold for the next two hours.

Mark skillfully got us to our position directly over where our gunships had to work below. You could see their anti-collision lights (An **aircraft anti-collision light** system can use one or more rotating beacons and/or strobe **lights** and have different intensities when compared to other **aircraft**), the red flashing beacon of both gunships far below us. Occasionally you saw the red ribbon of rain from the mini gun tracers, every fifth bullet is a tracer (**tracers** are bullets with a small pyrotechnic charge, making the bullet visible to the naked eye during daylight, and very bright during nighttime firing).

Mark informed the crew chief and gunner to arm the flares and begin deploying these artificial sun sticks into the darkness which completely surrounded our craft. The crew, aided by only handheld flashlights, began to set the timers so that the magnesium flare and chute were clear of the ship before exploding from their housing tube. They accomplished this by attaching the lanyard to a fixed point that enabled them to throw the flare while releasing the pin and arming the flare.

Picture these soldiers harnessed in monkey straps (Monkey strap is a long safety strap which allows the crew to move more freely in the cargo area) with cargo doors completely open, reaching down outside the ship, grabbing these flares in the darkness while thousands of feet in the air. This process would continue until the bins on both sides of the ship had been emptied. These flares are extremely bright and last much longer than a smaller artillery flare, aiding our gunships below to see the terrain and hopefully assisting them to avoid collisions with trees or ridge lines at their low altitudes.

There was one incident I recall where one of the gunships struck the top of a tree and shattered their chin bubble, the plexiglass just below and forward of the pilots' seats for viewing on such a night attack. Sometime inches make that thin difference between life and death.

During this usual procedure, something went wrong. We heard a pop, the sound muffled by the wearing of our helmets, followed instantly by a loud metallic bang which vibrated through the aircraft. Need I explain the 'Oh Shit' sensation?

The crew chief and gunner were yelling "holy shit." Mark was immediately on the intercom system yelling, "What the fuck happened?" One of the flares misfired early, driving the aluminum holding canister up into the sheet metal roof of the cargo compartment, penetrating the thin metal sheet and making a hole. Fortunately, the parachute and magnesium flare which were ejected prematurely just cleared the ship, not getting tangled or, worse, pulled back into the tail rotor.

Once this magnesium cocktail ignites, the heat is so intense it would burn immediately and instantly through the cargo floor to the UH-1H fuel cells of JP4 jet fuel…this material cannot be extinguished. If this had happened, we would have been the brightest object in the sky with all who could see our misfortune on Mother Earth now muttering in their best urban dictionary vernacular "EEEW," an artificial sun. Such events tragically have happened.

We landed after completing the mission and recorded the mandatory entries in the flight book of our damage that night. The Company maintenance sheet metal guys would be on this repair immediately, working to get this bird flying ASAP. Walking in the dark, we returned to our tin roof barracks and hit the rack, trying to go to sleep, knowing that shortly we must awake for the new 5 AM missions.

Mark taught me to take things in stride and try not to worry about the past because there will always be that next day to encounter.

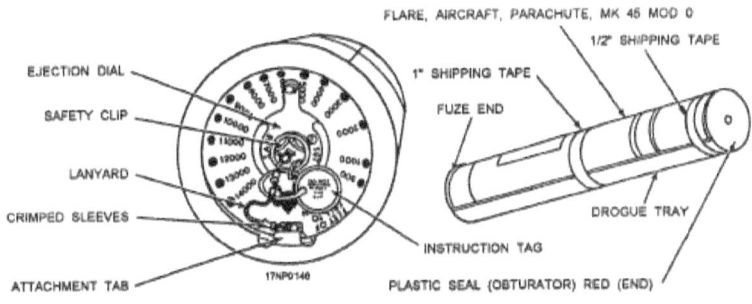

MK 24 Flare Settings

Flare Bucket

While writing these memoirs, I have chosen to add this accident report below of a tragic account of a flare mission which resulted in five fatalities. One of the pilot victims was a classmate and in the same platoon of 2nd WOC, Tommy L Kearsley, Army Aviator, and now another whose picture is displayed as guests enter my humble residence. Tommy is in the third row, third man from the left. I would engage anyone who now reads these stories to understand that during this war of ten plus years, thousands of aircrews would fly repetitive missions and as mentioned on the cover 'SOME HAD IT BETTER—SOME HAD IT WORSE."

CHECK RIDE

HELICOPTER UH-1H 68-16244

Information on U.S. Army helicopter UH-1H tail number 68-16244
The Army purchased this helicopter 0969
Total flight hours at this point: 00000474
Date: 05/04/1970
Incident number: 700504351ACD Accident case number: 700504351 Total loss or fatality Accident
Unit: A/101 AVN
The station for this helicopter was Camp Eagle in South Vietnam
Number killed in accident = 5. Injured = 0. Passengers = 1.
costing 622545
Original source(s) and document(s) from which the incident was created or updated: Defense Intelligence Agency Helicopter Loss database. Army Aviation Safety Center database. Also: OPERA (Operations Report.)
Loss to Inventory

CREW MEMBERS:

P CW2 KEARSLEY TOMMY L KIA
AC 1LT MATTINGLY LARRY FRANKLIN KIA
CE SGT AHLBERG THOMAS OLIVER KIA
G SGT TAYLOR RODNEY ALAN KIA
OB SGT AITKEN DEAN L KIA

THOMAS MCGURN

ACCIDENT SUMMARY:

ON 4 MAY 1970, UH-1H, SN 68-16244, FROM COMPANY A, 101ST AHB, WAS ASSIGNED FLARE STANDBY FOR NIGHT ILLUMINATION MISSIONS IN SUPPORT OF THE 1ST INFANTRY BDE, 101ST ABN DIV (AMBL). AT APPROXIMATELY 1930 HOURS COMPANY A OPERATIONS RECEIVED A MISSION REQUEST FROM THE S-3 AIR OF THE 1ST BDE TO PROVIDE ILLUMINATION FOR A PRACTICE RED ALERT AT FSB KATHRYN. AT APPROXIMATELY 1955 HOURS UH-1H SN 68-16244, COMANCHERO 20, DEPARTED COMPANY A HELIPAD ENROUTE TO FSB KATHRYN AND ARRIVED AT THAT LOCATION AT APPROXIMATELY 2020 HOURS. AT APPROXIMATELY 2030 HOURS TWO AH-1G AIRCRAFT FROM BATTERY B, 4/77TH ARTY DEPARTED CAMP EAGLE ENROUTE TO FSB KATHRYN FOR A PRACTICE RED ALERT. THE WING SHIP OF THE FLIGHT OF TWO AH-1G AIRCRAFT WAS 67-15620, CALL SIGN TORO 91D. TORO 91D WAS HAVING DIFFICULTY MAINTAINING VISUAL CONTACT WITH THE LEAD AIRCRAFT (CALL SIGN TORO 93) DUE TO THE FACT THAT TORO 93'S ANTI-COLLISION LIGHT AND AFT NAVIGATION LIGHTS WERE INOPERATIVE. TORO 93 INFORMED 91D THAT HE HAD HIM INSIGHT AND THAT THEY WOULD HEAD FROM THEIR PRESENT POSITION DIRECT TO THE FLARES OVER FSB KATHRYN AT 5500 FEET. THE FLIGHT OF TWO AH-1G AIRCRAFT ARRIVED ON STATION AT FSB KATHRYN AT APPROXIMATELY 2045 HOURS. TORO 93 MADE VISUAL CONTACT WITH THE FLARESHIP (COMANCHERO 20) AND CONTACTED JULIET 3 (FORWARD OBSERVER FOR C 1/501 INF) ON THE ARTILLERY FREQUENCY. AT THIS TIME THE FORWARD OBSERVER INDENTIFIED THE TARGET AND REQUESTED THE ARA MAKE THE FIRING PASSES FROM NORTH TO SOUTH. AFTER OBRAINING THE GENERAL LOCATION OF FRIENDLY ELEMENTS

IN THE VICINITY OF FSB KATHRYN, TORO 93 ELECTED TO MAKE HIS FIRING RUNS FROM EAST TO WEST TO PRECLUDE OVERFLYING THE FRINDLY POSITIONS AND SO INFORMED JULIET 3. DURING THIS PERIOD OF TIME COMANCHERO 20 WAS IN A LEFT HAND ORBIT AROUND FSB KATHRYN DROPING FLARES AS DIRECTED BY C 1/501ST COMPANY COMMANDER. COMANCHERO 20 THEN REQUESTED THAT C1/501ST COMPANY COMMANDER HAVE THE ARA AIRCRAFT COME UP ON UHF FREQUENCY E33. AT THIS TIME THE FLARES WERE IN SOME WAY GETTING TANGLED UP (REF CPT KOBINAR'S STATEMENT). SOON THERE AFTER COMANCHERO 20 COMPLETES DROPING THE FLARES FROM THE LEFT DOOR, HE THEN CALLED TORO 91D AND TOLD HIM THAT HE WAS GOING INTO A RIGHT HAND ORBIT. TORO 91D AT THIS TIME WAS FLYING THE RIGHT WING POSITION OF TORO 93. TORO 93 WAS MAKING HIS INITIAL FIRING RUN FROM EAST TO WEST WITH HIS WING MAN FOLLOWING HIM. TORO 91D DID NOT ACKNOWLEDGE THIS TRANSMISSION. FIVE MINUTES AND FORTY-FIVE SECONDS LATER THE GROUND ELEMENT, PLANTER 63, NOTIFIED COMANCHERO CONTROL THAT THERE WAS A POSSIBILITY A MIDAIR COLLISION HAD OCCURRED. AT THIS TIME THE BOARD FEELS THAT THE FOLLOWING EVENTS TOOK PLACE AFTER CONSIDERING THE EYEWITNESS STATEMENTS: THE UH-1H WAS ENGULFED BY FIRE APPARENTLY CAUSED BY A FLARE MALFUNCTION. AT THIS TIME THE UH-1H STARTED A DESCENT TO THE GROUND AND IN THE PROCESS MADE CONTACT WITH THE AH-1G. THIS CONTACT RESULTED IN THE LOSS OF THE UH-1H'S ROTOR SYSTEM WHILE DAMAGING THE AH-1G. THE AH-1G CAUGHT FIRE AND THE TWO AIRCRAFT SEPARATED, THE UH-1H FELL TO THE GROUND IN FLAMES WHILE THE AH-1G PROCEEDED IN WHAT APPEARED TO BE AN AUTOROTATIVE GLIDE AND CRASHED IN HEAVILY WOODED AND MOUNTAINOUS TERRAIN.

COMANCHERO CONTROL THEN LAUNCHED ANOTHER FLARESHIP TO FSB KATHRYN AND NOTIFIED HIGHER HEADQUARTERS OF THE POSSIBLE MIDAIR COLLISION. MAJOR PEASE, LATER TO BECOME PRESIDENT OF THE BOARD, TRAVELLED TO THE SITE AFTER BEING INFORMED OF THE POSSIBLE MIDAIR COLLISION, AND OBSERVED THREE SEPERATE FIRES ON THE GROUND ONE INVOLVING THREE SECONDARY EXPLOSIONS OVER A PERIOD OF FIVE MINUTES. THE FIRE, LATER DETERMINED TO BE THAT OF THE UH-1H, WAS DESCRIBED BY MAJOR PEASE AS CONTAINING GREEN FLAMES WITH THE SECONDARY EXPLOSIONS BEING WHITE.\\

This record was last updated on 05/25/1998
Date posted on this site: 09/23/2017

Copyright © 1998—2017 Vietnam Helicopter Pilots Association

12. Thee Bruce

Shortly after my arrival in LZ Betty, I moved once again and my new roommate was Bruce Britton whom I first met at Ft Wolters, Texas in 1968. We were both then assigned to 2nd WOC, training to be new pilots and were called green hats. WOC was the abbreviated acronym for Warrant Officer Course and there were nine Companies, from the 1st to the 9th. Each of these companies were separated by two weeks and there were four Platoons to each company. All had a distinctive colored baseball cap, clearly designating our competition of highly motivated future Army Aviators. The training units were separated by two weeks ensuring the unremitting surge of young Army Pilots needed as replacements for the ongoing war effort, "needs of the service."

Only those would graduate if they could endure one year of meticulous harassment specifically designed to make one attuned to attention to details, and the numerous flight check rides scrutinizing your learned abilities. This is what a war will do, very similar to the WWII production of pilots. The US Army needed pilots and it took just short of one very full year of constant training to become an Army Aviator with the reward of silver wings. So, the factory was in full tilt and we were all cognizant of where most of us were likely headed. Bruce had started with me here at 2nd WOC but would soon be set back due to a personal family emergency, he would arrive approximately one month behind me at Phan Thiet.

Now just before my arrival, the 192nd members had been living in tents surrounded by sandbags and were in the process of just completing new wood framed buildings covered with tin metal roofs for the officers. I was told a few weeks before I arrived at LZ Betty that a 122-millimeter rocket had scored a direct hit on one of the enlisted tents, killing three enlisted soldiers. Obviously on receiving such a dire communication, my expectations of future attacks were elevated and unfortunately fulfilled. During my entire obligatory detention at Phan Thiet, aka LZ Betty we were constantly mortared or rocketed with their first arriving announcements echoed in by that familiar deep thud type of explosion. Some of these incoming rounds were so close you could listen to the shrapnel hitting the tin roof over our heads, "too close".

The officer quarters were four separate wooden buildings, each dedicated to a section of the 192nd organization. My 1st Platoon quarters plus two other buildings one being the gun platoon and third comprised of HQs staff building formed a U-shaped pattern with a small courtyard where you could play volleyball. The fourth building was 2nd flight platoon which sat on a small hill to the west.

Bruce was a native of Nevada and although he didn't think so, he had that western slow way of speaking while being a smart ass. I was in country, meaning Vietnam, for about six weeks when Bruce and I became roommates, our room was second to last in our building, about ten feet wide and about twenty feet deep. The walls were made of pine wood with no ceiling, which allowed every dust storm to cover our beds in a clearly observable film of red dirt, a problem we would later solve with a parachute tightly nailed just below the rafters. Soldier inventiveness necessitated by environment.

Bruce was a true character and never thought to hold back comments on how he felt. I will always be so grateful that his humor made my time tolerable. He was a true brother from another mother, and to this day although we may only talk to each other twice a year, we both know we will remain comrades/friends till our last days. Later, we both would become PCs (pilots in Command) at the same time. This progression being dictated by older PCs who held us back a little due to the fact that not many new guys were being sent to our company for replenishment.

Sometimes you earned a call sign like Maverick or Goose in the "Top Gun" movie. I was given mine by Bruce after arriving back to our room, my flight mission complete, and now finding this slug "Bruce" busting into my canned food which had been sent in care packages from home for my personal comfort by my family. Care packages were always like Christmas Day and especially appreciated, there was no constant computer contact or phone banks which would arrive in the war theaters of the 1990s.

Setting the stage, I walked in after completion of my mission day tired, hungry, thirsty, and with a great desire for my custom stash of canned goods to now suddenly discover the big one **"THE BRUCE"**, sitting at our home built wooden two by four dining table feasting on my food. I started with my usual salutation, "Hey asshole, that's my stuff" which he instantly retorted, "You fucken **MOTHER!**" The word Mother being exaggerated with that stupid smug smile on his elongated face. We both laughed as I accepted his hunger needs and he later told all to now call me **MOTHER!** My call sign had arrived. Truth is, Bruce could have all our food, a small price to have him as a friend. Concerning my new call sign, to some outsider this may have sounded like OH! He takes care of his men, but it had nothing to do with this, you don't get to name yourself something cool. Bruce would have his day when late in the evening while the majority of the crews were sitting, drinking, and joking outside our humble home as our hero Bruce walked out of our room attired in his issued green army boxer shorts "very loose", his olive-green combat tee shirt also very loose and now heading directly to the piss tubes (an empty 155 shell carrier dug into the ground as a field expedient solution to a urinal).

Bruce a genuine personality, my opinion, who fit the description of Washington Irving's famous fictitious Ichabod Crane, (*He was tall, but exceedingly lank, with narrow shoulders, long arms and legs, hands that dangled a mile out of his sleeves, feet that might have served for shovels, and his whole frame most loosely hung together*). Bruce now standing there in that tall Greek statuesque pose, his inner attitude exposed, the low sun silhouetting this long-limbed thin man, skinny without any muscle definition, lengthy arms, slender pale legs, when someone yells like a shot from the grassy knoll **"Hey Atlas"**. The crowd roars with laughter and Bruce would be forever known as "ATLAS" until our time would come to leave country. At that time, we would be given a plaque of thanks from the company inscribed with our call signs, now immortalized, a small but well appreciated symbol of our time with this proud unit.

I have so many stories with Bruce woven into my permanent memory. It was Bruce my roommate who would pick me up after I was shot down. He would fly once with me on a flare mission acting as my copilot where we flew off into that darkness only wearing our boxer shorts and tee shirts after having been drinking wine the whole evening, not normal for me but we all lose our common sense at least once in our lives "right". We had taken off and climbed to maybe six or seven thousand feet and began to realize it was "freaken" cold as the crew chief and gunner greatly enjoyed our stupidity of flying in our underwear. Wish I had a picture, acknowledging their glee, with our jovial agreement.

Finishing our mission of emptying out all the flares over the action below, much needed for the gunships, we now decided to descend into the action ourselves, a decision after years of review was stupid with a big capital "S". Lesson learned here; even small amounts of alcohol have consequences. Flying low as red tracers and green tracers are sporadically flying through the night sky and to our surprise red tracers, which usually mean ours, all now coming upward?

As we started to descend over this small village on Highway One, tracers were still being shot skyward while we turned for home. Landing we placed our ship into its narrow revetment when we saw one of the gunships landing on the PSP airstrip twisting halfway around spinning on its skids, this ship had its tail rotor cable shot through by an enemy round, but the crew managed to skillfully land. Memorable night for them.

Another clear remembrance…Bruce and I were flying as copilots in separate aircraft supporting some ground units in the area at the southern end of the Le Hong Phong forest. It was the dry season and both ships had already landed, maybe fifteen minutes apart in a large open landing zone to deliver supplies. This would be a common occurrence when twenty plus 192nd birds were on daily missions in the AO (area of operations), we often could chat over the radio with our fellow first platoon aviators. It gave one that warm fuzzy that we were never alone. This landing area was level and the earth had been utterly grounded up by tanks and APC's, the remnants of such activity was the light-colored dust/powder that was almost the consistency of talcum. All the grunts were completely covered with this dirt blanket, like being hit by a clown from a circus stateside, who slaps someone with that large powder puff.

During our landing, the ships rotor wash (Air turbulence caused by a helicopter **rotor**) ensured our aircraft crew was also ensconced in this atomized dirt. This was the end of our mission day, along with Bruce's aircraft who also was completing their mission day. As both aircraft prepared to depart for home base, it was determined that Bruce's machine would takeoff first while we would wait until their dust cloud dissipated. During flight training at Ft Rucker, all pilots had received training in takeoffs and landings in brown out (dust/desert operations) and white outs (blowing snow operations). These training scenarios consisted mostly of what if applications since Ft. Rucker Ala., had neither large LZs filled with dust, or snow. The danger in these type of flight environments is loss of visual cues, especially the horizon. Such takeoffs required a true commitment to leaving, no hesitation as you increased power with a steep vertical climb to clear the dust cloud. Many helicopters have crashed in such conditions.

Both machines were at full power, rotors turning, and our aircraft was positioned about forty yards from Bruce's aircraft with our nose pointed directly at him as his aircraft began their departure. The vortices, cyclonic type air movement from Bruce's aircrafts rotor tips now sucked up the fine powder of the earth, causing an instant explosion type of cloud with its yellowish dirt now blocking out the blue sky. Seated in our aircraft, we strained to look for their aircraft emerging through the storm but only saw and ingested their dirt "thanks Bruce".

After a few seconds, you knew something had gone wrong, the cloud was not dissipating, and we never viewed Bruce's helicopter clear the cloud. We tried to call on the FM radio with no contact. A few more seconds and a blurred image of a helicopter emerged from the settling cloud of powder. The machine was running at full power, however it took our minds a few more seconds to realize that the ship's landing skids were completely collapsed, visibly crushed into the belly of the aircraft. Some would say hard landing, others a crash, I viewed it as a dent. No one was hurt, maybe only the PCs pride had been bruised and the very expected chance you would be the target of tonight's court side jabber.

Bruce was lucky, they had lost reference on takeoff and the PC chose to push the collective swiftly down, with the unexpected result. It was a little bit comical as the PC attempted to step out, assessing the damage and expecting a three-foot elevation only to immediately stubble into the dust, now level with his exit door. Again, aircraft have succumbed to such conditions often with fatal results. The Bruce lives and I'm good with that.

But my preeminent story of this war hero Bruce Britton was the time we were being heavily mortared with rounds so close there was impact pressure and shrapnel hitting our roof. We both woke from a deep sleep with echoes of "incoming" shouted by neighbors nearby and started a hasty course from our beds to the bomb shelter behind our building, constructed due to numerous mortar attacks.

Rushing out the door while turning left, trying to get back to the shelter, these incoming rounds were closer than they had ever been before. I was in the lead, Bruce directly behind me with his guiding hand on my upper back pushing me forward when what I would describe as a sixty-millimeter mortar round landed directly in front of me, maybe twenty yards away. You could feel the impact pressure and see those red-hot sparks of shrapnel flying upward just far enough to miss us both. I tried stopping, leaning back, shuffling my feet like some cartoon character as I then felt **"THEE** "Bruce placing both hands on my shoulders, grabbing my tee shirt, shielding himself while using me as his human defense, one of the bravest things he has ever done, and one I always bring to his attention to this date some forty-three years later. We did a quick turnaround swiftly returning to our room and lunged under our bunks to ride out this attack while our laughter of **"THEE Bruce's"** noble bravery was shared. I wish I had put him in for some kind of a medal, an award for conspicuous self-survival.

There would be many more stories with Bruce, and we would fortunately see each other several times during these next forty plus years. One of those friendships you take to heaven.

OUR HOOCH

Thee Bruce

13. Pilot in Command (PC)

Now after about six months of constantly flying missions in the Company, they had to replenish their pilot in command positions. This was on a rotational basis as the old crews departed Vietnam to return to a well-deserved going home after their time in this purgatory of the Vietnam conflict. Just as I communicate these stories, remember all pilots and crews who served have their tales. Imagine the magnitude of that equation of bad memories and laughter squared plus. Moving to the command seat was rewarding but also took each new PC to a level of responsibility where those he flew with or transported was a life or death possibility depending on the complexity of any assigned mission. Yes! I'm still twenty-one years of age and in charge of a five hundred-thousand-dollar machine in 1969 dollars, with a crew of three additional members who move at my command, but more like a request. I can feel the gray hairs hiding under my skin.

All the 1st Platoon PC's would now caucus to discuss the new candidates' abilities to handle their new positions. These senior pilots whose ages ranged from 21 to 27 had been flying with us for at least five months, they have taken us through every conceivable mission, and can evaluate our proficiency to function on any unusual situations we may have to encounter in our future flying. They were also our friends or teammates expecting us to finish our own journey as they now return home after passing the skills they perfected and practiced to us, the next generation of new PC's.

So together Bruce and I passed the endorsement process, whether we wanted to or not, and now would be assigned a ship and a crew. Each new PC would get his own ship (helicopter) and assigned crew chief and gunner. This did not exclude us flying with other crews if our ship was in maintenance, days off were rare. We all now began to discuss who we would like as our crew members. Crews were picked like children on a dirt ball field, who would finally be on their team.

Lastly left were a crew chief and gunner who I had flown with and respected their work ethic and dedication. They were both black Americans. Forgive me for not remembering their first names, I always referred to them by their last names. Moody was a crew chief and a very sharp focused individual who was meticulous in the care of his machine. I think Moody was from Michigan or some state in this area.

I once had an unfortunate experience with a short time crew chief who should have been more attentive to his machine during an event that could have killed all of us. I will never forget that lesson. While I could understand his lack of attention or concentration after twelve months and his looking toward to that homeward journey but could not excuse his loss of focus which hastened those aircraft problems. Moody was my choice, proven through my previous flying experiences with him, a true soldier, and a man who would contribute to all of us getting home, not a difficult decision for me.

Now I had to choose a door gunner, a position where he assisted the crew chief and maintained our fire power which consisted of two M60, 7.62 caliber machine guns. These machine guns, if not properly maintained, could be fickle and jam or single fire…not welcome functions on any mission. When these guns worked properly, they were impressive and easily kept the enemies' heads down, giving us a huge edge in survivability. With five to six hundred rounds per-minute going down range, it always made you warm and contented inside, a form of dessert, that comfort of covering fire.

The gunner usually seated himself on the right side of the aircraft while the crew chief was positioned on the left. It was in these strategic positions they could best use covering suppressive machine gun fire to ensure our touch down and take offs while in a hostile environment. My Gunner's last name was Dudley, again I apologize for not knowing his first name, but I knew that Dudley would be my gunner. I had also observed Dudley's actions before becoming a PC. Dudley would often stay at the aircraft at the end of the day, meticulously cleaning and maintaining the weapons while assisting the crew chief with the everyday maintenance of the ship, a very hard-working man. I think Dudley was from the Chicago area. He was just a good man, I almost lost him on a future mission. There was no hesitation about choosing my crew. I knew my best chances of getting home was with these two men, both great Americans as I remember them, then and always.

Now flying combat missions involved the whole crew, the team, four sets of eyes, four brains concentrated on the mission as briefed, and constant communications through the internal communications radio. Flying constantly into the jungle, the landing zones were sometimes so tight and constricting that you would never have made it to the ground without a blade strike or something more disastrous occurring without the crew chief and door gunner directing you to move left, right, down, up or not to touch down due to stumps or other potential damaging material which could puncture the ships thin bottom skin (something I once did by a hidden stump in the high grass).

I think every helicopter pilot in Vietnam has clipped some tree branches or leaves unintentionally during their tour, but without a good crew, those incidents could turn fatal, resulting in the ship crashing and rolling left or right or worse. Without Moody and Dudley, my chances of going home would have been significantly reduced. They gave me a feeling that every time we took off we were at our peak in our performance. I trusted them to do their jobs but always knew they exceeded this effort and I thanked them for that. I was proud of my crew and will always reminisce of Moody and Dudley for their exceptional soldier skills and combat brothers always.

14. Long Day But!

This date we were assigned the Admin route beyond the mountains to the West, most likely supporting MACV locations within II Corps. MACV was an acronym for Military Assistance Command Vietnam which controlled all operations and troops in Vietnam. Under this command were Special Forces like the Green Berets who coordinated with South Vietnamese Army and local village and town representatives.

All the routes west of the mountains were flown over heavy triple canopy jungle and were situated anywhere from twenty to thirty miles from the Cambodian border. This was the major entry sanctuary for the North Vietnam Army to attack South Vietnam. Principle landing locations were the villages of Gia Nghia, Bao Loc, Da Lat and Dak Som with the occasional lone outpost maybe five miles from the border, these were way out there the tip of the spear, made me truly appreciate our home base with that South China Sea view; location, location, location.

The mission was basically a bus run to deliver anything, soldiers, people, food, ammo, mail, everything needed to support each individual mission. Imagine the surprise of some newly reassigned soldier being delivered to one of these jungle Oasis's, knowing he is trapped there for the next year. This was a fairly casual assignment with a big plus that normally we could shut down for lunch at one of the better MACV locations where for some reason their food was excellent. Like eating out at a restaurant rather than enduring our home chow which could be gruesome sometimes. Tasty meals always appreciated. These missions usually meant a long-extended day, maybe ten plus flying hours and even at our closest position, we were still a minimum of a one-hour flight away from home base and would still have to cross back over the mountains, something we would rather not do at night. Often in the evening, cumulus clouds would cover the peaks and high spots of the mountain ridges, better to use the daylight. Cumulus clouds are the big puffy ones that look like cotton, they have a vertical development with defined edges but if the sun goes down, you don't want to run into one with limited IFR capability (Instrument Flight Rules). At the end of the day, the crew is tired, hungry, knowing we missed supper again, and will roll into base just before sunset hopefully. This is the yin and yang of flying.

There was a communication network in place for limited flight following a rudimentary system to attempt tracking our location throughout the day. Often our Company would have resourced this communication system to reach out and amend ongoing missions, if necessary. As a result, flight following personnel could attempt radio contact and would then have the unpleasant task of informing us of a last-minute addition to our mission. Such notification "untimely as it is" always resulted in placing a burden on an already exhausted crew.

On this day, near the end of this mission day, we would receive such an urgent request. The message ordered us to proceed to an outpost to pick up and transport a seriously injured civilian. Now normally a US Army Medical Evacuation (MEDEVAC) ship might be assigned but their assets were often stretched out to the maximum and often over worked with US troops being their priority, "as it should be", those guys were truly noble.

We worked out a course heading and developed a quick plan of action followed with a mixed intercommunication crew conversation consisting of some pretty foul language and "why us". The "why us" turned out to be we were the nearest asset available and capable of helping this poor SOB. Fortuitously, we had just enough fuel to get to this outpost and on load the injured party then fly directly back to LZ Betty, which had the best and nearest medical facility to stabilize him.

As we approached the outpost, the landing area was clear and made of PSP metal pad, (**Pierced Steel Planking, used to make temporary airfields**) soldiers are already waiting with the injured subject to be transported. Upon landing, the crew aids in placing this wounded man on the cargo floor (a hard aluminum/magnesium metal) making him as comfortable as possible.

I now received the brief on our unfortunate medical passenger. He was a Montagnard, a native of the Central Highlands, kind of like an American Indian. He had an abdominal wound with some protruding intestines. Just knowing his ailment sent that uncomfortable feeling through my body, a sympathetic reaction we all have felt sometime in our lives. I was told it was from a hand grenade explosion but with no explanation of which side inflicted such an injury. Obviously for us to assist this man he must be an ally of some kind.

We placed him in the middle of the cargo floor, strapping him securely in case of any unforeseen turbulence in route, the crew closed our cargo doors. He would be warmer, that should help. Our guest was shirtless, wearing shorts and sandals, and clutching a hastily made compression bandage around his mid-section. You could read the pain on his face, he did not make a sound, just sat in a squatting position as we took off for an hour plus trip to home base at LZ Betty.

Now airborne for about fifteen minutes, we noticed our passenger was also clenching a metallic object the size of a grapefruit. The crew chief moved in to examine what this poor injured man had in his hand and reported in his astonished voice over the intercom that our boy was holding a 105 Howitzer round fuse. This is the piece that sits on top of a

105 round to dial in the pre-explosion height after being fired, I'm pretty sure it has some kind of primer.

The troops on the ground most likely had given this to our patient to help keep his mind distracted from his serious injury, "shiny metal always works with Indians, remember Manhattan." My crew and I may have chosen a different trinket, maybe a tuna can or spoon, anything without the explosive factor. Too late now, we decided to let it ride and would inform the medical team to be prepared to separate him from this explosive device upon landing. Long trip, but this man never whimpered, never complained, and barely moved for just over one hour. You had to admire his courage, his presence, and we were rooting for his recovery, imagine holding your intestines for any amount of time.

The ship was now five minutes out as we contacted LZ Betty tower informing them our first destination on the airfield would be the medical landing pad at the north end of the field, they acknowledged and facilitated our approach. As we touched down, the medical aids were waiting with a stretcher to move the patient as quickly and comfortably as possible to the triage room. Our job was over as we took off toward the POL refuel pads, topped off, then back to our Company landing area to place the ship in its protective revetment until the next sunrise. The sun had dipped below the horizon twenty minutes ago, there might be some food left in the mess hall, they always had leftovers for us, it usually was so pitiful but still welcome and an end of a very long day of flying was over. This was one of the many times we never found out the result of our efforts, but I choose to believe this man lived after seeing his enduring stamina.

CHECK RIDE

MONTAGNARD VILLAGE MISSION

I have chosen to add another mission which was so amazing to me that I shall never forget the experience. Once again, we were supporting MACV on the route's west of the Mountains. While refueling at Da Lat, a beautiful city with mostly French Colonial architecture structures, a gem in the Central Highlands, we now received instructions to proceed to a location to transport two unknown civilians. Our passengers arrived, they were cordial and spoke English with that distinct accent one would expect. A man and a woman, both maybe in their late thirties or early forties, dressed in khakis, looking excited for their adventure to begin. We onloaded their equipment which consisted largely of medical supplies and learned we would be tracking out to a remote Montagnard village along a riverbank and they assured us all would be fine, "clearly they haven't heard about the war".

After picking up our paxs and while receiving their destination coordinates, we plotted a course for the LZ. It was immediately evident that this place was just that, a coordinate, no villages, no outpost, nothing but that deep jungle green coloring on our maps and some unattractive high mountains on our planned route. This mission most certainly must have a high priority because the two passengers were doctors from some Nordic Country, missionaries maybe. This maybe was fine with them and I respected their humanity, but we would be arriving to this unknown location in one of the most hated aerial devices of the North Vietnamese Army outside of US bombers. This was a single ship operation to within ten miles of the Cambodian Border with no escort at max fuel range, "but, hey, we had a good lunch".

The flight was not routine, how could it be? Since we left DaLat, we had been flying over jungle for forty-five minutes straight when we arrived at the location. I will tell you this, the US Army makes you learn how to read maps, catching every detail, contours, rivers, streams, mountain cuts, cliffs, saddles, ridges and grid coordinates, and this is where it pays off.

Below us was a very small village on a muddy riverbank. The river was about thirty feet wide with some wood canoes on the bank. We did a low fly over of this area to complete an aerial reconnaissance of our touch down location, making sure we didn't encounter obstacles on approach and just gave this raw landing spot a final check. On short final, almost at touch down and now taking in the entire village scene, "I was in a time warp". The village was small with raised wood huts, actually everything was wood. The village had some fortification emplacements of large wooden poles mixed with bamboo stakes surrounding the entire hamlet. I was reminded of my grade school history classes with pictures of wooden forts on the early American frontier, or a National Geographic magazine with some Amazonian tribe, how lucky to live this exploration.

The residents had all come forward to watch the show as we landed with that impressive effect of the swirling wind generated by our rotor downwash, they were barely dressed, the women were naked from the waist up, children completely naked, the men just wearing old shorts or loin cloths but their expressions were priceless, all of them smiling. Children had that ecstatic gaze fixed on their soft faces like Santa Claus had arrived. This scene was so rustic, so far removed from anything I had ever experienced.

I was lost in the moment, but quickly realized we were deep in enemy territory and here we are the enemy, we needed to get going and not just for fuel consumption reasons. Our passengers unloaded their supplies with the assistance of the locals, such a happy people, very appreciative, collectively. With a wave of thanks, which we acknowledged with a small salute were off backtracking to civilization where we belong. The sky felt so good.

At some point, I had been given a Montagnard brass bracelet, a very simple piece of rolled brass with some small marks carved into the metal which I believe I gave to Kerry my daughter, perhaps she still has it.

Another long day, new challenges and now back home to the familiar vestiges of the modern world. Food, shower, drink followed by a night's sleep on my mattress.

Similar to this but smaller and on a hillside

The term Montagnard means "mountain people" in French and is a carryover from the French colonial period in Vietnam.

15. Sniffer Missions

One of the more unusual missions we had to fly in Vietnam was fittingly called a sniffer mission. All through my enrollment in flight school, there was not one mention of such a mission, "I guess they didn't want to scare us off too early." Never tell volunteers the reason why no one else wants to do this, "Sssh, it's a secret."

This sensing device was invented by the General Electric Company during WWII to detect ammonia a byproduct of humans, animals, fire's etc. I would have the opportunity to fly several of these specialized missions during my tour in Vietnam. This particular assignment I found enjoyable, despite the fact that we were flying low, slow, and extremely vulnerable. The mission was designed to reconnoiter areas most human eyes will never see. Sniffer missions were usually pre-dawn takeoffs, for reasons I will further explain as I take you through a summary of some of these missions, which I can recollect for the obvious reasons.

As a pre-planned mission, it would be posted late the night before on the Platoon mission board outside the Platoon commander's room. That was how we were initially informed of our next day's assignment. A sniffer mission involved equipment which had to be especially installed on the helicopter, I have attached a photo (below) to show the mechanics of this process.

    The crew was up early, still dark as we prepped for the flight. The equipment was installed and we, the crew, reviewed this day's mission planned route of flight with the enlisted technician who operates this sensing device from his seat in the cargo area. There was no enthusiasm from this young soldier, you can read that childlike expression on his face. He would rather be in a safe barracks still sleeping rather than flying over the menacing enemy jungle in a single ship at tree top level. These missions were always conducted over the triple canopy mountain jungle areas because the data interpretation from the machine would be useless on the farming flats where there were an abundance of inhabitants and domesticated farm animals, making its effectiveness much less objective.

    We took off with a chase ship and a pair of gunships to ensure our potential rescue, mostly this set up just gives one that guise/good feeling option of a Team. The chase ship and guns would stay at altitude in the vicinity of our planned track to reduce our total noise signature. We departed the runway, heading north out over the mud flats of Phan Thiet harbor. The tide was out and below the citizens were on the mud flats taking their morning bathroom break, using the South China Sea tide as the big flush. I'm sure this process goes back centuries. The city has those mixed fragrances I will forever associate with the third world that blend of charcoal cooking, fish byproducts produced in the city, and tidal mud flats. I will never forget this sour perfume as we now turned left north west into the emerald green mountains to complete this day's mission.

    We were now flying deep in those mountains, rugged jungle, truly a lost world. On this one day, I was flying with my co-pilot Thomason a good stick and gentleman. The sun was just breaking the horizon with wind none-existent, a plus for the sniffer machine. No air disturbances and maybe another reason why these are early morning missions. The crew chief and gunner were keenly aware with weapons pointed down, and out, fingers on the trigger, ready to rock. Everything we now flew over was a potential target.

We flew at tree top level doing forty knots (46MPH) an extremely easy target for any enemy. Lonely down here, hell, they could hit us with their shoes. Later while writing this, I would discover through some research on the internet that the enemy had actually absorbed their consequences of firing at helicopters on such missions. They had learned the hard lessons of giving up their locations which would bring the pain of immediate multiple airstrikes, gunships, and artillery barrages.

The sniffer was supposed to pick up chemical compounds from campfires or groups of humans (don't rule out monkey's or herds of elephants) while the soldier/tech plots possible enemy locations on his grid map for higher command interpretation back at base. The jungle was so beautiful this time of day, everything so tranquil, the temperature was cool and humid with a low light ground fog over the streams, exotic birds dashing at their top speed while monkeys in the tree tops observed, and the sun was streaming through the lush vegetation like a masters painting, "paradise looking down". Vacation views...not many people will ever get a chance like ours to look at this striking beautiful world of nature, there are some good things about flying.

Working our way up a stream maybe ten feet off the water, low grass pulling away from the stream with huge jungle trees hanging over the path of this water, our speed had dropped below ten knots. We have gone to a visual search (very risky) playing LOH (a light small observation helicopter) when we spotted a campfire smoldering next to a small lean-to shelter and some hanging clothes on the edge of the stream. We were not alone here. Apparently, we must have roused someone up and relayed our sighting to the chase ship, just in case this turned bad. The green light for the crew to go guns hot was always on the moment we were flying at tree top level and on mission. No reason to lose that precious reaction time with audible request at this juncture, plus this whole area was a free fire zone.

Now moving forward, the terrain turned into a bowl shape as the earth rose to an area surrounded by a ridge. I have to tell you all who now read this memory that this is the only time I have ever felt my hair stand up on end, giving me an out of body feeling with an internal message of, "Tom you don't belong here," but I'm the PC and in charge of this ship and this mission, but this impulse was instantaneous and a little overpowering, "very eerie."

On the intercom, I let Thomason know of my apprehension, looking for a little crew participation and support here, and I was surprised Thomason stated he felt the same exact sensation. What are the chances of this happening to both pilots? Maybe I had turned that corner and now being on the downhill side of getting home chose to imagine this threat, but it was something else which both of us experienced. I have never in my entire life been so influenced that something was telling us this was a merciless place and my decision was quickly absorbed to not enter this area.

Our technical sergeant operating the equipment agreed, he never really wanted to be here anyway, probably forced into this assignment for the needs of the Army. We decided to just post it on the report and maybe sometime in the near future this area would be scheduled for a B-52 bombing run (the military solution, kill everything). These hills were swarming with the enemy troops always moving south on multiple routes through Cambodia by the thousands secreting themselves in the deep cover of the jungle, until repositioning forward for attacks. It was the ultimate game of hide and seek with the loser possibly facing their untimely death.

During my tour, as I have mentioned, I completed several of these missions and at the completion of another mission I would fly directly over a substantial enemy camp. This took place in an area we called

CHECK RIDE

the "toilet bowl." A mountainous area due north west of LZ Betty, the same area I flew that tragic night mission when we lost my friend John Wright. The mountain tops peaked out at about thirty-nine hundred feet and were very heavily covered by triple canopy jungle. On this day we were near the completion of the morning's sniffer mission and flying back to our home base of LZ Betty (Phan Thiet), but still working the sniffer mission box.

Still flying that very low profile and reaching the base of the toilet bowl area when an opening appeared in the jungle foliage, unbelievable you instantly could see down into this void surrounded by high trees where shockingly about two dozen NVA soldiers were sitting around a small camp fire, weapons by their sides, "damn!" We were just as surprised as they must have been. Of course, they could hear the UH-1 helicopter with its distinctive blade popping from miles away in the quiet early morning, but maybe they were so used to this now common sound and constant harassment that they felt protected by so much jungle.

As we passed overhead, I could see them quickly looking up, startled expressions on their faces as they now scrambled for their guns while running away from the clearing and heading for deep cover in the jungle. Some were diving headfirst for their lives, knowing what was surely to follow. The crew chief was on the ball "as always" immediately popping a smoke grenade to mark the location as we informed the gun ships to proceed to our location and that we had clearly marked this area with yellow smoke, radioing the gunships "enemy troops in the open."

Two gunships have a considerable amount of fire power and would literally soak this area with rockets and mini-gun fire. I would not want to be in that neighborhood when the gunships arrived in the next two minutes. There would be future occasions where I could clearly see the enemy again and always be pleased to not have that future knowledge of when?

After such mission, we would return to base going immediately to hot refuel as always, you never park an empty ship. You don't know when you may have to saddle up, and fuel is king. Perhaps if it was still early and we could sneak back to the mess hall and get some eggs. I know I have made claims about our mess facility being less than stellar, but they always tried to take care of us knowing flying schedules were unpredictable, "God Bless those Guys".

The day was not over, too early to waste useful blade time. So, we would be reassigned to some type of supply missions for the rest of the day. Supporting the numerous units in our AO (area of operation), soldiers somewhere in need of water, food, and ammunition. Everyday seemed different and we were flying, life was good.

People sniffer — an electronic device that detected smells that came from humans, tied to urine and sweat. This could detect enemy personnel in hidden positions, often in jungles.

16. USS St. Paul, Last of WWII Gun Cruisers

It was an Army Pilot's dream come true we were assigned to fly out and land on the USS St. Paul, a World War II Gun Cruiser and the last of its class still serving. It was off the coast of Phan Thiet, Vietnam in 1969. Only a battleship is superior to this type of ship which has now participated in its third war since it was commissioned in 1943. Writing these stories forty plus years later has given me the chance, no the advantage, of researching this great ship's history through the internet, to completely examine how unique the USS St. Paul was, a true piece of American Naval History. At the time of this mission, the USS St. Paul had been assigned to patrol the Vietnam coast and we often would observe her silhouetted against that split between the sea and horizon, a very impressive mental postcard of American presence. *("A ship is called a she because there is always a great deal of bustle around her; there is usually a gang of men about; she has a waist and stays; it takes a lot of paint to keep her good-looking; it is not the initial expense that breaks you, it is the upkeep; she can be all decked out; it takes an experienced man to handle her correctly; and without a man at the helm, she is absolutely uncontrollable. She shows her topsides, hides her bottom and, when coming into port, always heads for the buoys.")* During her last few months, she cruised back and forth, North to South using her eight-inch guns to engage enemy targets on the coast in our area. In a previous account, I related how I was on a Charlie Charlie (command & control) mission where we exploited the power of this war machine and called in a direct fire mission from this magnificent ship.

Hard to recall but I think our flight this day had an early start and I say "ours" because we were a crew and always a team, amalgamated by the closeness of our tasks. I only wish my very seasoned memory would allow me to recollect that everyday crew chatter, that combination of comical statements, helpful suggestions, and just some common-sense tips that we all need throughout our lives. I think maybe this day we had been assigned to a log mission (logistics), resupply runs for those units in the field, ferrying the normal daily needs of combat units. These missions are usually quick and start from a PSP pad on the west side of the airfield. Supplies are positioned in stacks designated for each individual unit and each stack has its weight calculated to be within the lift limits of the UH-1H helicopter, "hopefully". Sometime there will be a significant number of sortie's, perhaps ten or more, and sometime there may be less than three. So, the potential to end early is always a welcome surprise.

Our first mission had ended as we refueled and checked back with

CHECK RIDE

192nd flight operations to inquire if this day will be the exception with an early day off. Flight operations contacted us and requested myself as the PC to shut down and immediately check in personally to the flight operations building for a new assignment, always a crap shoot. Too few times I have left flight operations with a smile on my face. They had now given us a primo mission, one the whole crew will especially remember, and enjoy. We have been tasked to carry the US mail to the USS St. Paul, a weekly assignment and the only way that this ships compliment of combat sailors while on patrol can received those personal touches from their loved ones back home.

I briefed the crew on frequencies, positions, and procedures that I had received directly from flight operations personnel. Operations had instructed us to make radio contact with the St. Paul and to be prepared to land on the fantail of the gun cruiser. "Fantail", what the "F" is a fantail? I'm an Army pilot, never been on any ship other than the Mayflower replica in NYC harbor as a child with my parents, "that's it", please explain? So, reduced to simple Army instructions it is the rear of this mighty ship and the rear triple eight-inch gun turret will be repositioned appropriately, pointed out to sea to increase our landing zone, simple, right?

The Huey is no small machine. Its length with rotor turning or stopped is fifty-seven feet, it stands over fourteen feet high, and is just over nine feet wide, and we will be landing this machine on a moving landing pad, trying not to think like a kamikaze. Today vs yesterday such a mission could not have been completed without a certificate of pilot training and qualification on ship borne landings and takeoffs from a designated US Army Aviation Course. Now travel back again to these forty plus years and all it required was, "here get this done we have complete confidence in your ability—go team go."

Airborne en-route to the USS St. Paul flying over the South China Sea, looking forward you see her. She emerged at first small, like those plastic Revel models I made when I was a child with my friend Rick. But approaching fast, she grew as we sped ahead. In my mind, I was constantly reassuring myself, "I can do this." Radio contact was now established on the USS St. Paul's assigned frequency as we requested permission to land, and now receiving this authorization to land from this magnificent grey vessel, although delivered in a much more formal manner than we were familiar with.

Maneuvering our helicopter to position ourselves at a forty-five-degree approach angle, aiming for the rear (fantail) of this moving grey behemoth. Closing in I quickly realized my concentration was on two points, one being the deck landing pad and the other on the ship's rapid movement through the ocean with the sea water churning and creating a white foam, all being rushed rearward, very distracting imagery. This was one of those learning lessons literally "on the fly" as I comprehended how to adjust my full attention to the landing area only, this would automatically engage my trained Army Aviator brain to direct my input control actions and steer our ship to land safely while still moving with the forward speed of the ship "on the fantail". Flying everyday will always improve one's skills as corroborated by this successful landing and completion of this story.

Onboard after shutting down, we had been invited to lunch by the Captain and crew, and this was then followed with a quick tour of one of our Countries premier seaborne moving embassies. I was in awe of seeing my machine sitting below these triple eight-inch gun turrets on the back (fantail) of this war machine. We were now greeted by a Naval Chief Warrant Officer who would escort the crew to the Officers Mess for lunch. The welcoming we received for just delivering the mail was unexpected and jovial. These sailors had gone beyond welcome, more Americans to be admired.

Lunching now in the Officer's Mess (Dining Facility) with the Captain and his staff of officers. The Captain of the USS St. Paul was a

CHECK RIDE

full bird Colonel, one step below Admiral and he was gracious to his guests. The crew and I were seated at a long table complete with white linen table cover, matching linen napkins, real silverware, and real glasses for drinks, not those Army brown composite cups and dishes, a scene right out of the movies. There were actual waiters dressed in waist length white serving jackets with naval emblems and gold braid. These sailors/waiters were Philippine, and this was their full-time assignment.

Having often been served by ourselves at our meager mess facility where I once found a dead rat in the Kool-Aid, this experience I would compare to a five-star NYC Restaurant (note to self—Go Navy). The meal was composed of a fresh salad, soup, homemade hot bread (heaven), real butter, choice of drink (non-alcoholic), and the main course Beef Stroganoff, I should say that once more Beef Stroganoff, real food, "Beef Freaken Stroganoff". Now for the past five to six months eating mostly out of a can you can imagine we were blown away only to then have coffee and dessert, real ice cream (thank you Jesus). Concluding the meal phase, we then were given a tour of the ship, clean bunks, flush toilets, and no dirt. Quickly moving now, our time to leave was approaching fast.

Cranking up on the fantail (practically sailors now), the Chief Petty Officer ran to our machine with two very large paper bags. This was not their outgoing mail which we had already stored for the return flight, but fresh baked doughnuts' and rolls for our fellow soldiers back at Phan Thiet. I guess everyone had heard of Army food, often the target of jokes by professional comedians which got us much long-awaited sympathy from our brothers on the St. Paul. Always greatly appreciated as we cranked up and began our take off from the "still" moving ship.

I once again was concentrating on some part of the ship until we were completely clear of all structures now zooming low over the Ocean as we started to climb. Heading homeward and will be feet dry (over land) in fifteen minutes, flying back to reality and our normal missions.

First stop to POL for fuel then onto placing the bird into its revetment until the next day's mission. The crew chief and gunner were in charge of the gifts from the St Paul, they knew better than me who really needed a lift me up. I proceeded to flight operations to give some sincere thanks for those responsible in their intelligible selection of my crew for this mission. This was a once in a lifetime experience, I even kept the takeoff manifest with the St. Paul's letterhead and believe I have hid it away with some old memory stuff, but this one is so easy to bring back. Vietnam wasn't all doom and gloom. You had to enjoy what you were given and laugh when you can, life lessons.

YEA!! I found it after an entire day of doing my rain man act among those lost scrolls of the basement area filled with mites, cobwebs and dust bunnies that hide like they had an existence, the USS St Paul's mail manifest. What humored me most was the date. It is probably in my last two weeks of flying from LZ Betty before I depart to Home. I would have been the senior and most seasoned Pilot in Command in the first Platoon on 16 May 1970. Perhaps that was their (operations) intent to have a proficient pilot complete this unusual mission. Or, it was a gift, presenting a once in a lifetime experience to a departing friend. I choose both and now am proud to display my memento.

CHECK RIDE

Mail manifest

Ship's Fantail.
/noun/

1. The back end of the ship on the hangar deck level is known as the fantail.

17. Another Night Flare Mission, only as PC

(COLD SWEAT)

These remembrances are not in order, except for co-pilot to pilot in command, so here is my first incident of scaring myself and questioning my abilities, was I really ready for this?

I was flying all day with my co-pilot Norman Niswonger, yes that's a real name and his call sign was NORM only because he probably went through so much shit as a child with his last name that we decided to call him Norm. Norm was a great guy, always happy, never serious, and a pleasure to fly with. He enjoyed beer, smoking, laughter, and always had a smile. I say this because after the war Norm came and visited me in Yonkers NY. He was from Ohio originally and had a young daughter. Norm's wife had passed away before he was in Vietnam and his daughter was his life. Unfortunately, Norm was killed in a civilian flying accident while crop dusting in an OH-13 model the very next year after his visit. I have no doubt, nor does Bruce, that a lifelong connection would have been maintained with Norm. I think of him often.

We had been on a mission all day and I'm pretty sure it was the dry season where multiple brush fires were smoldering, mostly these controlled burns had been initiated by farmers who were burning off their rice fields throughout our entire AO (area of operation), a seasonal event.

This large-scale conflagration contributed to a permanent smoke haze with that burning grass smell, even at some altitude. Not too disturbing during a daytime flight but multiply this factor of less visibility and its increased effects during night operations results in what seems to be a tenfold loss of horizontal visibility during the blackness of the Vietnam night.

Flying in such conditions is hard to explain but place yourself in a completely dark room with furniture positioned that is unfamiliar to you and try walking without striking those obstacles, or bumping into a wall, or even tripping because you cannot see your feet. Now add altitude, speed, and panic.

After arriving back at LZ Betty, we learned that my crew, including Norm, will be the stand-by flare ship this night. "Oh goodie." The crew chief and gunner started the evening post flight inspections then continued right on with the special installation of the flare buckets, "no rest for these guys." Again, the buckets are attached to the outside of the cargo area and each one holds a dozen plus MK-24 magnesium flares. I was by now familiar of all the dangers of such a mission as I have previously mentioned during one of my co-pilot experiences. Looking back on this time with my aged perspective now matured to a more common-sense solution, I would love to review the standby mission crew selection memo.

Like any stand-by mission it was hit or miss, you may get a full night sleep or hear that now all too audible launch alarm for mission on! Waking up from a deep sleep, trying to get dressed while grabbing your flight gear, then running in the dark to your ship as trepidation tingles one's mind. Sometimes ships will launch in support for LRRP (long range recon patrol) pick up or, like this one, a flare support for our Tiger Shark gunships on their way to assist a ground unit under fire in the bush.

Once more we had drawn the low card and have been ordered to launch. Typically, we departed after a hasty briefing, obtaining a grid

CHECK RIDE

coordinate to locate our intended support area and quickly compute our on-course heading as we took off into the black night. The dark was even more obscured by the same smoke haze from those farmers' brush fires, a very uncomfortable feeling as we climbed to altitude. We practically lost all reference with the below ground features as we passed through another thousand feet, continuing to ascend.

On this night as a fairly new PC (pilot in command), I was flying in the right seat where our instruments display is larger and packaged for IFR flying (IFR means instrument flight rules), hoping it gives me an extra edge. Norm was now in the left seat, eyes wide open, brand new experience for him, this was his first night mission. Flight school doesn't prepare you for the reality of an unprepared mission.

We climbed between six to eight thousand feet with the dim red flashing anti-collision lights below of the gunships as they started their strafing runs in support of the infantry unit in distress. We occasionally caught sight of the red line of mini-gun tracers' rounds pulsing through the dark. Now positioned and deploying MK24 flares to illuminate the area below, it was helping to highlight the ground definition beneath for our gunships, all seemed well.

We were flying pretty much on instruments now, absolutely no hint of a horizon. My crew were trying to always assist, constantly giving some guidance to orientate my course with what they were observing directly below. Peeking out the front windscreen, I could see nothing but milky night, no reference to the ground, and than the mind starts its tricks with the body, as gravity, speed, and banking turns work to rapidly induce vertigo, most always caused by your semicircular canals in the ear which respond slowly to such movements.

This sucks! I was uncomfortable with the experience. I felt like I was sitting sideways trying to cogitate/think it through. It overwhelms the senses immediately as you feel like you're sitting ninety degrees to your position, as your own body attempts to fight off these erroneous feelings, knowing that the aircraft instruments correctly reflect the situation, but helpless to stop the brain's mistakes. An instant cold sweat springs on you, running through the entire body, a physiological stress of potential aftermaths induced by the pilot's best friend and enemy "adrenaline" as you continue to struggle trying desperately to overcome this awareness of panic.

Knowing the numerous stories of ships which had been destroyed, resulting in a total loss of the crew because of night spatial disorientation, I was trying to resist this onset of cold sweat but rapidly losing while watching my gauges and seeing my airspeed increase then decrease as I chased the attitude indicator, attempting to reign it in as it dipped to the color black below the horizon plane and climbed to the color white twenty degrees above the horizontal plane, fluctuating thirty degrees plus or minus (view photo attached of attitude indicator).

I needed a break to get control of this and informed Norm I had vertigo; Norm instantly comprehends. With this announcement, the crew chief and gunner knew we are in trouble and both can only sit there riding it out with their own minds racing with their possible doom. Norm came to my rescue and responded, "I have the controls." The relief of hearing those words had an immediate effect, a rush of hope helping me believe we as a crew can get control of the problem. But within just maybe two seconds, Norm's voice, now cracking, related he also had vertigo and our porpoising (the nose of the aircraft moving steeply up then steeply down like the mammals in the sea with about the same speed) which has now actually increased from 40 knots of airspeed to 130 plus, back and forth, we were in deep trouble of losing the ship.

Without accurate airspeed, our instruments will be totally faulty, driving us into bigger and quicker mistakes. I took the controls back and told Norm, "I have it." The fear was building but I still had to fly so I instantaneously lowered the collective, centered the cyclic, and placed the bird into a significant descent while holding a fixed heading to now stabilize our ship as I brought the airspeed back under control. The maneuver had the immediate effect I was looking for and that sweet sensation of control had returned.

We dropped a few thousand feet and were able to make out some ground definition, helping to insure control of our flight. We continued the mission at this lower altitude as the crew chief and gunner reset the MK24 timing fuses to adjust for a lower opening. The flares could still be deployed for our gunships; however, their overhead burn time would be less, but the safety of our ship was now secured with some ground reference finally in view.

UH1H Cockpit 1970

Attitude Indicator

When you graduate Army flight school, they give you a "tactical instrument rating" knowing it's just enough to maybe get you out of trouble. Accidents related to spatial disorientation were very common during the Vietnam War. I had a flight school friend who was killed in Vietnam by just such a situation where his ship went inverted at night with no possibility of recovery, his name was Rocky D. Armstead, a good old Texas boy. He was the best pilot in our flight class and a friend. His picture is on my foyer wall with my flight class where he remains forever young.

Landing back at LZ Betty, Norm commented that he was scared shitless there for a few minutes. He was impressed how our descent stabilized the ship. He said in his earnest smiling way he had learned something. Imagine, I was only five to six months ahead of Norm out of flight school and he thought I knew everything. I couldn't hold back but let Norm know immediately that I also was scared shitless, with another capital "S" in bold print with a large font.

Of course, this experience was a first time for me also, if that wasn't obvious enough. This exchange was followed by some very nervous laughter, that kind of chuckle that falsely reinforces the imperviousness of youth as we both completely understood at any time we might again be tasked a similar mission, only to face this same scenario once more, only now armed with the knowledge of a possible recovery procedure forever stored from this combat experience. However, walking silently to my hooch, my subconscious was vociferously shrieking, "HOLY SHIT! I'm fucking alive."

The true heroes that night were my crew chief and gunner who never said a word, but their presence onboard was my obligation to get them home safe, how brave of them to put their lives on the line in the hands of others. It is humbling to think of that courage. We didn't pick our missions in Vietnam, they were assigned, and only our obligation made us blindly complete them, some training obviously was on the job.

I would fly many more missions with Norm and so would Bruce, he learned like we did and became an excellent PC. I miss this character who I was lucky to have known and called friend.

Very old pilot adage; "Experience is something you get after you needed it." Most likely spoken by Orville to Wilbur **on the hallowed sands of Kitty Hawk.**

18. Lift Mission, Lead Ship

Seniority brings responsibility and this comes in the form of more complicated missions with increased accountability. Growing up fast now, serious stuff, I had to climb into my big boy pants. I was twenty-one years old, which I have mentioned several times, but seriously looking forward to twenty-two, hoping age brings wisdom, "HA!"

The 192nd Assault Aviation Company had a compliment of about two hundred men and was broken down into Platoons within the Company. First and second Platoons were flight assets, with pilots and crews flying slicks (troop carrying helicopters). Each Platoon had ten helicopters. There was also a gunship Platoon which had approximately six gunships with their crews, call sign "Tiger Sharks."

A crew is always two pilots, crew chief, and a door gunner. Four united souls with the common core goal of surviving every day. In addition, there was a maintenance Platoon, responsible for the daily upkeep of all our ships, and to make sure they will be flyable twenty-four seven. Rarely did we go under eighty percent of flyable aircraft.

Finally, there is the Headquarters Platoon that encompasses the command staff with the commander, usually a major. Also, within the HQ's Platoon would be the mess staff and all non-aviation Officers and enlisted men, the main gears running the Company. If you look at our military, they have honed management and organizational capabilities effectively, with procedures established rather well, and many of these soldiers who leave our services bring multiple skills perfected through experience back to everyday jobs at home.

Larger missions often will come down the pipe from Battalion after being regurgitated from higher Brigade and the Division, everyone has their pennyworths of input. They know the who, what, when, where and why of any mission, and guide its ultimate goal to completion.

On the evening before this lift, I was told I would be lead aircraft on the first lift, followed by a second lift within minutes. A new step up for me leading, nervous to do it right. "Yikes!" lead PC is in charge of this flight. When conducting these more complex missions all the pilots gathered at the operations room for a detailed briefing. Maps were posted with SP (starting point), PZ (pickup zone), IP (initial point), and LZ (landing zone) all plotted on each PCs grid map. Frequencies were assigned and route sequencing was established with exact times displayed on the briefing board. Each lift would have six ships carrying six fully loaded combat troops and four crew members per aircraft, about sixty souls flying into a hot LZ (landing zone). Between the two flights, seventy-two combat troops will be plunged into battle.

This was an area where it was highly suspected that the enemy were residing. It would be a coordinated insertion and timing would be critical because of the inclusion of artillery support (also named steel rain) scheduled to be continuously poured into the LZ right up to thirty seconds before the first lift touched down. This artillery effort would be awesome, and the LZ would be pulverized.

Visualize the reality of dozens of 105 artillery Howitzers which would be firing a twenty-pound warhead up to 11,600 meters or about seven miles away from our LZ, directed with precision and accuracy onto the designated enemy area. Our Tiger sharks/gunships and crews would also be present to cover the grunts as they first seized this new

CHECK RIDE

ground/LZ. The Tiger Sharks would remain on station for the next hour as a precaution, just in case the area erupted into an instant fire fight.

Obviously with this close artillery prep, higher command suspected the enemy was nearby. All PCs were busy plotting SP, PZ, IP, and LZ with on course headings and approach headings to avoid the gun target line (An imaginary straight line from gun to target. Also called GTL), of the artillery rounds being rained down into the LZ. "Very important." You do not accidently want to pass through this unseen line in the sky. Every PC is a potential flight leader, a necessity built in for redundancy in our business, we are all leaders, lessons of history. You must be prepared for the unexpected but pray it will not happen.

The briefing also had the command and control pilots who would coordinate any last-minute changes along with the field unit commander who would be seated onboard while flying at altitude in the command control ship to incorporate any last-minute changes if necessary. I had been on several of these but this would be my first time as the lead ship taking sixty men into the unknown, I was excited to have been chosen, but deep inside I hoped I didn't, "F it UP."

Once more the early morning sun was just breaking over the South China Sea, another spectacular day, with that early morning sparkle of gold on the ocean's surface, nice living in the high rent district. Time had now arrived to begin as the twelve slicks started their engines in unison at the exact briefing time, it was a beautiful site. All these helicopter turbine engines cranking, with that low familiar whistle sound now blended with the oily smell of the burning JP4 kerosene fuel. This combination always gets your blood stirring to just be a part of the operation, an inner growl of the warrior.

Pressure waves generated by the twelve helicopter blades turning with their unique sound of wop wop wop pulsing, this is the exclusive characteristic of the Huey Helicopter as it penetrates our helmets.

Temporarily you forget its war. The gunships are also running as all our war wagons go through the mandatory startup checks and radio checks.

My crew chief and gunner are well prepared, guns cleaned, and ammo in place. If for some reason our ship ever goes down, the crew chief and gunner have the most fire power on the ship, in addition beside their M60 machine gun, each have their individual M16 automatic rifle. If we are fortunate to survive the crash or a forced landing, their new job becomes securing the area.

As the radio checks were completed, I checked in with operations on the Company frequency for any last-minute changes, and now informed them that flight one is up. The second serial/flight will be a few minutes behind us and were on different frequencies to prevent possible confusion of flights talking over each other. Radio discipline is so important, idle or stupid chatter could override important information, this experience you learned in your first days of flying.

Flight one was up as we relocated to the west side to pick up these young soldiers, they are trained well as they placed themselves in perfect order to load the ships. All were burdened with large rucks, loaded to the max with mostly ammo and food, they know better than anyone what it takes to survive. Americans should see this up close sometime, the dedication, bravery, and pride in their unit as these young soldiers once again prepared to place their lives in danger for the ideology of governments.

Remember Armies just don't go where they want; they are instructed by politicians. We were right on schedule, the party was about to launch northwest to foothills, just the noise from all these machines vibrated the ground. It was exciting to participate in this. I was stoked!

Clearing the airstrip at LZ Betty, we climbed in staggered right formation. Ships were barely two rotor blade lengths away from the other. You could see our soldier's legs hanging out from the open cargo doors on our sister ships, their weapons at the ready, and they could see each other with only their individual thoughts of what will happen in the next twenty minutes.

As we passed over the large city of Phan Thiet, you wondered what all those people were thinking as they watched us fly toward the hills, knowing that such a large formation of ships with our cargo of soldiers would probably be engaging some of their countrymen somewhere, "soon". I actually then thought they don't care, because for thousands of years these people learned to survive for themselves.

The course we selected was plotted on our map as we checked our time to meet the schedule, we hit the IP (Initial Point) as planned. Good news, right on time as Charlie Charlie gives us a wind update and confirmed our landing direction would be as briefed for the LZ.

The wind direction is power, you don't want a tail wind pushing your ship faster on landing. Now stealing the power, you need to touch down. Just the turbulence created by six helicopters fully loaded landing in a small confined area can cause havoc. Approaching the LZ, we were flying a parallel course before we began our turn to the final approach path. You could see that the artillery prep had begun with multiple explosions, steel rain impacting the landing zone, this would continue until we were on short final.

Flight one is good, our formation was tight as we turned onto final about four clicks out of the LZ, now directly ahead with the artillery rounds still tearing it up, these are no small explosions. Closing fast now on short final and the artillery fire hasn't stopped. I was a little worried here to how close this artillery fire was, and I just started to initiate a check fire call to halt this barrage as the artillery fire than stopped. This was confirmed by the CC ship who radioed check fire was in place.

The dust had started to die down as I gave the order to the flight for guns to go hot. Six outboard M60 machine guns now tore up the tree line on both sides of the LZ, tracers flying everywhere. The LZ had a small right to left slope, but manageable. I told my copilot to fly as far forward toward the tree line as he could to ensure all our flight had room to land, this technique we did learn in flight school.

As we touched down, skidding forward on the ground, the troops dismounted, each one choosing his place to defend as they moved to cover, their job now was to cover us as we departed. This LZ was right at the base of the mountains, a regular path of the enemy troops. This assault could go sideways quickly.

All my six chicks' radio they are up, we go out as a flight, all six taking off in unison now empty with power to spare as we followed a plotted path back to Phan-Thiet, LZ Betty. Flight two would be just a minute behind us but would not come in hot while our troops were already in place to avoid fratricide and the LZ had been secured by the first lift.

These soldiers probably would be in the field two weeks or more and future resupply missions would be the responsibility of the 192nd. Resupply missions always made you feel good about bringing water, food, ammo to these guys. They certainly deserve praise. Back at LZ Betty, my six ships formed a trail formation as we all headed to the hot refuel pads. This was the POL area where you landed to a small cement pad and with the engine still running the crew chief grounded the aircraft to prevent static discharge from starting a fire. The chief took the fuel nozzle, similar to a gas station except larger, then opened the fuel cap and at high pressure filled the fuel tanks.

I was always amazed that we never caught fire while the blades were turning, engine running, and raw fuel was all over the place. After refueling, we checked in with flight operations, parking the aircraft in their revetments until later when they would be dispersed to various other missions. There are so many small little details which I forgot, but that's what happens over forty years.

Enemy forces then withdrew both to the Le Hong Phong and primarily (according to intelligence) to the mountain range north of Song

Mao—fundamentally due north of our last contact with them. The acting squadron commander then made what I consider to be a tactical error and turned Troop C into an airmobile infantry company and we air assaulted about 115 troopers to blocking positions in the mountains. (Researched from a combat report webpage)

19. Why

Looking back through these events, I realize just how many night missions I had to fly during my tenure at LZ Betty and ponder how I was so fortunate to return home. It is truly amazing if you can begin to understand the enormity of Army Aviation. There were twenty slicks (UH-1H helicopters) in our Company with six gunships and our Company was part of the 1st Aviation Brigade, 10th group. Counting our sister Companies, of which I think we had four Companies in our Battalion. These Companies were spread throughout II Corp Vietnam, in places like Dong Ba Thin, home of the 92nd Assault Helicopter Company, Ban Me Thout home of the 155th Assault Helicopter Company, and so forth. Every Company always had a ship prepared for night standby to include gunships, ready every night, 365 days to be geared up for an immediate launch for some unit who felt they had no other options available. So, to singularly think that I was being selected as a night crew member unfairly was a little self-regarding. We all were just objects of scheduling.

Night missions were precarious with all the varied flight conditions that may be encountered; no moon light, high cloud cover, low clouds, sporadic rain showers, no ground lights, smoke haze, ground fog, and these are just the in-route conditions. To order a nighttime launch of any aircraft was a serious decision and hopefully would acquire some meticulous analysis of the potential mission to be conducted by knowledgeable aviation officer's familiar with such hazards. In reality, the mission was forwarded to our Company operations from upper command, probably with little or no thought except that US soldiers need assistance NOW!

Comparison wise, Vietnam helicopter operations would be considered like flying the cloth biplanes of WW1 compared to the OIF level of sophistication of 2003.

An unscheduled mission was called into our Platoon commander by phone, a hardline connection directly to his room from operations. The dark had arrived, and the majority of pilots were still just hanging out, sitting in their individual cheap nylon multi-colored beach chairs purchased during some past mission to a large outpost which had a PX, (Post Exchange) most likely Phan Rang airbase. The Air force had everything, but hey, they had to fly north **"ugh"**.

Those were the days one could imbibe legally, and many pilots treasured their nightly relaxation session, all except those now on standby status. My platoon leader approached, truthfully, I do not have a clue, since the day I had arrived, standby mission notification usually started by an alert siren for the gunships sounding the first hint of trouble. The Boss started talking, telling me as standby ship I had been assigned to an emergency resupply mission to a unit in dire need (is there any other kind of need at night), get my crew, and report to flight ops. First, I grasped my copilot out of his beach chair and instructed him to notify the crew chief and door gunner, and to assemble at the ship. I can never say enough about my crew and all the crews, they just always braved these dangerous types of missions and made them so considerably more stress-free.

Operations had given me the brief on who, what, when, where, and why

CHECK RIDE

of this mission. Some unit of ground troops deep in the Jungle mountains needs supplies, ammo, food, water so urgently that this night mission was approved. The grid location was located and plotted on my map. It was at the south end of a mountain range we called the toilet bowl area, a very heavy triple canopy jungle zone with that deep, deep dark green looking vegetation you see during the day missions.

Whoever tagged this moniker to this area was a genius. The ground unit had just blown out a jungle LZ with det cord, a plastic explosive which they wrapped around the trees to enable our landing, clearing a hole so that we could get their supplies to their location. Built by ground troops who "guestimate" the size of a helicopter and who have been humping all day in the humid jungle heat; always expect the worst?

Best part of this brief was we would be a single ship, no flares, no guns, no chase ship, WTF! This area was known for harboring large enemy concentrations. Once, while I was a copilot flying low down a wild mountain river near this location, we came across what I would describe as the George Washington Bridge of bamboo most likely constructed by CHARLIE, "impressive", towers on opposite ends with suspension ropes made of jungle vines over a raging mountain river.

Leaving flight operations with my less than detailed instructions, I proceeded to the bird to brief the crew. I couldn't wait for those puzzled faces confronting me. Cranked up, we moved to the West side of the airfield in the dark, landing at the logistics pad to onload the emergency supplies. The ship was crammed, the entire cargo compartment from top to bottom with boxes of food, crates of mortar rounds, water, and small arms rounds, we were maxed out. Setting the aircraft interior panel lights to as low as possible, a trick learned early on that aids our much-needed night vision, we completed a lift off power check to ensure we can take-off. It was time to go, using every trick I had learned to conserve power. Airborne heading westbound and climbing into another obscure Vietnam night, leaving the ground lights and security of LZ Betty behind with nothing but blackness ahead, and those mountains rising "Gods Walls". Not one grumble from my guys as we shifted into gear getting airborne and, on our way,, then once on track it was a joke fest most likely to ease the concerns we all had, that nervous type of prattle concentrated on WHY us.

Steady on course and reaching our cruising altitude, we shifted to all business, dialing in the supported unit on our FM radio while simultaneously contacting the Bird Dog fixed wing on UHF radio who would aid us to the unit's location. We attempted to use our FM homing to also assist us to the unit's location, giving us an inbound course on the homing instrument needle, if it was functioning, while they transmitted to us, anything for an edge.

ADF Compass (A **radio direction finder** (RDF) is a device for **finding** the **direction**, or bearing, to a **radio** source)

Radio Magnetic Indicator (RMI)

Arriving over this unit, there were not many reference points. It was extremely dark out here, but we could now pick up their strobe light which punched its concentrated rays vertically through the canopy. We were now preparing to descend into the tiniest LZ ever made, "I think", completely surrounded by high jungle trees, those really big suckers, made bigger by night and imagination.

Hovering over the LZ, our power was nearly at maximum available with this full load, while we maneuvered to fit, no squeeze, into this hell hole of an LZ. Slowly inching our way down, watching our power gauge move to its maximum. My crew was now constantly talking of obstacles, tree limbs on every side of our helicopter, the crew knew their jobs and I had every confidence in their abilities. Our leading edge of the rotor blades were nicking the leaves, making that familiar slapping sound. Often when working such areas in the jungle, our blades would strike some small branches and you could get away with it, but that usually was during the daytime. "This sucks! A phrase I would too often use"

It was impossible. We could not come straight down, the LZ was too small and some of the overhanging trees were preventing a clear vertical drop. The crew chief and gunner communicated for me to move back and down, constantly yelling, "Keep the tail straight, don't move left, or don't move right." The copilot was concentrating on the front and right side, apprising me of any spare room to maneuver. The aircraft lights caused a multitude of visual shadows with the surrounding trees, very visually eerie.

My control movements felt so exaggerated, but I barely moved my hand on the cyclic. We had to do something out of the box here, we had to back down into and under the canopy to land, which by the grace of God (truly) our crew accomplished. We now sat wobbly on a pad/landing area made in the dark, running at full throttle and looking into those jungle shadows caused by our lights as the soldiers urgently rushed the ship like it was their last hope. They quickly cleared the cargo compartment, similar to those worker ants you see in jungle documentaries.

The voice of the ground unit commander was ecstatic for this delivery and you could feel the oscillation of his voice as he annunciated his sincere grateful thank you. He has kept a promise to his men; "always take care of your men."

One more little obstacle...we now had to reverse this puzzle of an LZ to depart since our rotor system was under the tree canopy, something made so much easier with the now endless power provided from our Lycoming L-13 engine as we departed with our unburdened machine.

L-13 Trubo shaft engine tested to be 99.9% reliable. We don't talk about the other 0.1%

UH1 Engine

The return flight was relaxed we could now see the lights of LZ Betty and the City of Phan Thiet in the distance. It was a fairly short mission but overwhelmingly received. We facilitated our guys, ensuring their security with those supplies, our reward. POL was closed after dark and the aircraft should be refueled by truck as we landed and lined up for our revetment, sliding the bird with only three feet on either side as the machine resisted with buffeting due to the rotor downwash.

Finally touching down inside the shelter and shutting down, we said goodnight to the chief and gunner as I headed to flight Ops for a quick debrief. You just cannot accomplish this type of mission without the crew sparking on all cylinders. Upon a safe return a sound sleep follows before tomorrow's missions.

Forty-four years after this mission, this was the most difficult LZ I have ever landed in and I have been to more than my share of hell holes in multiple countries. There is no better feeling than to know you were there for our fellow soldiers and that was the true mission of Army Aviation.

Partial LZ Betty

20. Lift Extraction, Mountain Style

Checking the mission board the night before "routine", I saw that our ship had been assigned to an extraction of an infantry unit in the mountains northwest of Song Mao. They, Flight Operations, had assigned four ships and a briefing would be held at flight operations after breakfast, not an early mission, heaven, sleep till seven. We had a rare sit down to a table breakfast with eggs to order, hash browns, bacon, toast, coffee or tea, French toast maybe, or some substitute, back in the normal Army.

Another beautiful day looking out to that million-dollar view of the ocean with the sampan's dotting the sea fishing as we walked back to the Hooch. We were gathering our equipment and conducting a leisurely start of this day. I was at that stage where flying was a job, almost commonplace, just do as they say and come home later to relax, feeling like you could do this forever.

At the briefing at 0900 hours in operations, our ship would be chalk 4 (**chalk** is a group of soldiers that deploy from a single aircraft), the last ship during a two-ship rotational route to exfil (exfiltration or **exfil**, is the process of removing personnel when it is considered imperative that they be immediately relocated) these young men to their well-deserved oasis of a normal living environment. After what seemed as standard discussion on how the lift would be conducted, frequencies, grid coordinates, paxs (passengers) we gathered at our ship with excess time to brief the crew chief and gunner on times, location, expected loads with drop off locations, as preparations were completed.

This was an extraction, taking troops out of a known location where maybe they had been observed by the enemy the past few days. The NVA viewed this as an opportunity to bag a helicopter, payback for the agony Army helicopters had inflicted upon them during this long war. The crew was dressed in full battle paraphernalia, chicken plates, survival vests, sidearm, armored side plates for pilots, and armored seat cushions for the chief/gunner.

Once again that time had arrived as we started up our machines with that all familiar whine of the turbines turning scented with the kerosene fumes from their exhaust as the ship vibrated to the increased rotation of those two big rotor blades, music to Army Aviator crews. Communication checks were completed as the lead ship gave the order to line up in trail formation as we proceeded to then take off northbound toward Song Mao airfield where we would stage for this mission in the mountains.

Flying at our standard cross-country altitude of fifteen hundred feet, a height determined to be safely out of small arms range and to avoid those errant rounds sent skyward by numerous VC and NVA who infest the brush forest below. Aligning with Highway One and following closely, we finally made our turn forty-five degrees to the northwest to prepare to land at Song Mao Airfield. Landing in trail, similar to geese landing on a lake, all the aircraft began to shut down, waiting for the mission to commence.

The unit being extracted was located about a fifteen-minute flight up the valley in the mountains, consequently we didn't refuel to minimize our weight, better to be lighter when picking up, especially in a PZ surrounded by jungle trees, aka obstacles.

Grasp that this is a unit about the size of half a company, seventy plus

CHECK RIDE

men, infantry, in the mountain jungle where they have set up a perimeter for defense which they now had to disassemble before leaving. These men had been there several days and their usual day probably started with a healthy breakfast of canned C-rations heated over a homemade stove GI style with a small piece of C-4 explosive acting like sterno to heat their meal...powdered coffee or chocolate maybe followed up with crackers and peanut butter, finally cigarettes, no shower, no water buffalo (M112 500 Gallon tanks that are towed behind, usually, a five-ton truck), no toilets, only self-reliance. These men had looked forward to this time through all their exhaustion, patrols, nightly guard duty, watching the sunlight sneak below the trees for the past several days, revealing those dark shadows of possible enemy ogres where the eyes struggled to place a shape to the changing twilight scenes for their nighttime detection.

Upon departing, they will attempt to leave nothing that CHARLIE can use. So much equipment to export, light machine guns, mortars either 60mm or 82mm, boxes of ammunition, C-rations, concertina wire, and all their individual ruck sacks filled with essentials of field living. These men are regular infantry having helmets, flak vests, and personal weapons. Their time in this location was certainly perilous but they were always in the arch range of heavy friendly artillery if they came under attack. The unit commander must organize his men into chalks, taking into consideration weight of equipment and the final cover for the last few lifts as they exited this makeshift forest fortification.

Sitting static at the airfield in anticipation of getting this mission done, the noon sun had arrived and moved forward to past midday. With this the heat had reached its peak, stealing that edge of power our turbines needed in the mountains. Finally, we got the word to crank up to fetch these boys out of the wilderness. We departed Song Mao airfield in a trail formation with a five-minute separation. The PZ was a one ship hole so we flew a chain to rotate this unit out.

Our ship was on final with my copilot flying, this was one of his training periods as it was my obligation to prepare him for his term as PC in the next few months. The PZ (pickup zone) was surrounded by incredibly high trees and it sloped right to left with one small area that one ship could touch down.

Empty, we had plenty of power coming in, the UH-1H helicopter has a max gross weight of 9500 pounds. With just the crew being combat equipped with full fuel we would be flying at approximately 7500 pounds, giving us a useful load of about 2000 pounds. This lifting capacity would be increased by another 600 pounds because of fuel expended arriving for the mission. Picking up our first load of six combat paxs fully loaded with some extra equipment maybe totaled 2000 pounds as they jumped onboard, sitting on the cleared cargo floor.

Unstrapped for departure, that fixed look of liberation was cemented on their smiling faces. The chief and gunner gave us the ready call as I viewed the power instruments, the midday temperature was taking its toll or maybe the engine was just getting worn down like all of us as we lifted off with power now just under max. The CP lifted the helicopter straight up, then inched forward toward the tops of the trees at seventy plus feet. Our power was good as we rolled over the treetops, picking up clean air to accelerate clear of the PZ. Flying down the mountain, we approached an easy landing at Song Mao, touching down as these grateful men debarked.

We took off once more, climbing back up to the mountain ridge for the last pick up. CP was still flying; I remember being mentored like this just several months in the past. On short final again, we would be the very last ship in and out. This is the moment you worry, that time before mission completion, that period your mind wanders to the most unpleasant thoughts, are the enemy watching or worse, waiting. Crew chief and gunner had weapons on the ready as the last group of souls started to climb in with all their equipment, including a freaken mortar tube with base. An even bigger surprise was an additional two troops, now a total of eight fully loaded combat soldiers who had maintained a protective perimeter cover until this last ride out.

We could squeeze them in on the open cargo floor, and we had to jam them in. You do not leave two soldiers alone on the ground trying to coordinate their own pickup by radio. We completed a quick power check necessitated by soldiers caring for soldiers. It was going to be close as the ship started the slow ascent to the treetops. The crew chief and gunner had begun to sweep the PZ with covering fire for our final departure, except as we reached this high hover out of ground effect, our RPM began to bleed off, the engine and rotor needles slipping into the yellow caution area of the gauge. We won't clear the trees; we don't have the power.

I have every confidence in the copilot as I tell him to ease back down into the PZ, which helps to restore our power setting into the green. Weight is the problem here; we have to shed some and fast. Glued to the instrument console in the UH-1H is a power placard which coverts torque power of .25 % torque will equal 100 pounds of actual weight. We don't need much of an edge for a quick fix so I quickly tell the chief to dump excess weight, he is on it, yelling at the infantry guys to throw the mortar base off along with any unneeded equipment.

These seasoned soldiers instantly realized as we slipped backward that something was amiss. Hearing the chief, they were more than complying with his request as the mortar base along with excess ammo and some additional non-essential equipment was promptly cast out over the side like a sinking ship into the vacant dirt below battle field litter. It didn't take much just that little bit of power as we once again lifted up and forward as the power reached its maximum right at the top of the trees, the copilot began his slow rotation forward over the trees and down the slope to gain the much-needed relative airflow and airspeed. We were out.

That friendly hot air breeze of freedom immediately filled the ship and the soldiers' faces broke out in uncontrolled smiles. Flying down the valley, my copilot apologized for the first attempt which I immediately let him know was not his fault, that his execution was perfect and for him to now remember that dropping weight can give him that power punch he may need someday when he too will be a PC and facing a similar event.

Landing with the last chalk at Song Mao, the troops slid out as we touched down, most likely off to the showers, I presumed, as we positioned over to the POL for hot refuel and our flight home.

Not much flight time this day, however textbook execution under very hot conditions and lessons for my future replacement as a PC. These are tricks I learned on my own CP flights from my friends long gone home and for me to now pass on. Losing power on takeoffs was a common occurrence due to weight and heat in this country but would turn critical in situations like this one. I felt much older.

21. Holy Crap, that's what they look like

We were working the mountains northwest of Song Mao on this flight day. Song Mao is a small village about thirty-five minutes by helicopter from LZ Betty. Today if you look at Google maps, it is named Phan Ly Chan and there will be a bus stop icon showing Ga Song Mao. As mentioned in several of these recollections, it was a very active area. At this location, there was a basic uncontrolled airstrip with an American fortification at the northwest end, again look at Google Maps. In this location you can still see the outline of the airstrip and even the disturbed area where the fire base was once located, scars of conflict.

The fire base is one of many used for artillery support to provide overlapping fire for our troops who routinely worked this area. The village has dirt streets still to this date, with small tin roofed shacks as homes for the locals. Near the airfield was a garbage dump with refuse, mostly from the artillery base. The airfield had a POL point (which stands for petroleum, oil and lubricants) where we often refueled during daytime missions, a necessity of extended flight operations in this area.

Every time we flew in and shut down, you could see the local kids barefoot walking through the garbage pile, covered with flies, looking for anything they could use, old food, metal, cans, and bottles. Sometimes the local kids were selling Saigon Cokes, a local beverage which tasted like Coca Cola and was served in reused coke bottles, probably from the garbage heap. They would walk from ship to ship begging for C-Rations scraps and some crew members, the more courageous, would buy these drinks as the kids pretended to pop open the bottle.

    I always viewed it as liquid dysentery in a bottle, but we rarely were back home for lunch and often quite hungry and thirsty. In Vietnam, I learned one thing well, never trust the water. Time was on the bug's side and we would all get sick during our tour; it was almost impossible not to.

Song Mao was a hot spot of enemy activity and I had flown numerous missions from this location, both night and day. The firebase was often under attack and on some future occasion I would be sent on an expedient mission after a major attack which I will narrate in another mission story. There are hills and mountains west and northwest of this village and judging from all the missions we conducted here, the enemy was well established. Many times, we were static at this hot, dry, exposed airfield waiting for instructions from the unit we were supporting. This was a kick-off point toward trouble.

    As a jeep approached with some liaison officer assigned to the unit we were now supporting. They pulled up with a map in hand, an obvious hint that a mission had started to take shape. We were now informed that we would be extracting a LRRP team (Long Range Reconnaissance Patrol) from the mountains on a scheduled pickup, thank God! Coordination is everything, no mission was ever a walk in the park here. We had been designated to pick up a five-man team from the mountains above Song Mao where they had been for several days, constantly moving, hiding, watching, and reporting on enemy activity.

    Communication was usually first, knowing the proper frequencies and backup frequencies to contact this team and having constant contact

CHECK RIDE

with the Bird Dog, a fixed wing aircraft whose pilot would help guide us into the PZ on the first try, which is critical. Grid coordinates were next to plan our approaches and departure from these mountain valleys and to visualize prominent features to aid our selected flight paths while tracking on our maps. Using grid coordinates, we consistently used at least an eight-digit numerical, to confirm and pinpoint the recon teams exact PZ.

These missions involved a total of five aircraft. Two UH-1H slicks, one pickup and one chase, two UH-1C gunships for immediate cover if the mission had problems, and finally overhead would be a fixed wing OV-1 Bird Dog (High wing Cessna) for directions. These Bird Dogs were usually single piloted aircraft used to monitor units deep in the mountains and to relay any radio messages back to their unit. This was a normal mission for this type of aircraft, and they tracked these teams every day, making them very familiar with any particular team.

When things went wrong for the recon guys, the first to know would be the Bird Dog who at altitude would have the best possibility of enhanced communication to aid them. They were literary an airborne radio relay station.

As time to go approached, the gun ships would come from LZ Betty where they were continuously on call. Gun guys fly less and have more down time, but they also are often more exposed to enemy ground fire. On this lift, I would be the PC of the pickup ship, while the second UH-1 loitered in the vicinity, again in case things went sideways.

Lift off was early afternoon this day, our maps were out and marked as we began our climb through the foothills toward the mountain area. It was the dry season and in some places, you could see the tiny paths which weaved through the jungle. These paths appeared to look like the deer runs you see back at home. We would go in as a single ship armed with two M-60 machine guns and whatever personal weapons we had amassed onboard. Usually these PZs were exceptionally tight, just barely larger than our rotor diameter. You had to descend below the jungle trees, which always seemed immeasurably larger as you lowered the aircraft.

We linked up with the Bird Dog who started giving us instructions on our inbound course. The Bird Dog may be at three thousand feet above us and had the whole picture flying a slow orbit as we moved up a steep valley listening to his instructions. As we approached the PZ (Pick Up zone), we decreased the aircrafts speed as Bird Dog called out one click (Click on the grid is one thousand meters or about thirty-three hundred feet). I alerted the crew chief and gunner, guns on standby (as if this is still needed to be said) as the helicopter was maneuvered to carefully descend into this hell hole of a landing area, while hoping the right people were still there.

We didn't request a smoke grenade if we could avoid it. We didn't want CHARLIE seeing anything if they were in or near the area. Below was brush grass with a small left to right slope. We quickly determined that a touch down could be made. The jungle was right on us as the LRRP team rushed from the deep forest refuge, their guns covering down as they limberly moved to board as quickly as possible, always appreciated. This was the critical moment, the transition point where both the team and the ship were most vulnerable.

The moment the team was secure onboard, the crew chief and gunner went hot with their machine guns, ripping up everything on both sides of this jungle PZ, firing five hundred rounds per minute; I love that sound. Immediately the LRRP team also joined in firing on both sides, expending every round from their weapons. You never can have enough bullets flying on departure.

We wanted to do everything to make sure we left this place. As we checked our power gauges, we started our slow climb straight up to clear these obstacles of nature. Moving over the obstacles, we slowly moved the cyclic forward to gain clean air and airspeed, announcing on the radio that we were in the clear.

CHECK RIDE

Having just cleared the PZ, we were tracking just over a mountain trail, still very low, when we saw a full squad of about ten or more NVA (North Vietnamese Army Regulars) making their way toward the PZ we had just left. I didn't have to instruct the crew chief. He had already thrown a smoke grenade to mark the area as I informed the gunships of this sighting. The guns arrived in minutes and worked over the area with mini-guns and rocket fire, saturating the area with a deluge of metallic rain.

This was a holy crap moment. We looked down, they looked up, they were dressed in pith helmets, weapons slung forward and NVA uniforms (I have attached a picture of what they were wearing), right out of North Vietnam. They had to hear us, the UH-1H has a very loud and distinctive blade noise, no mistaking US Army in the area. I think they were also caught off guard, I could see them look up and my guess was they were smart enough to know what was about to come their way. My admiration for our LRRP teams keeps growing, knowing they creep around these woods which are swarming with the enemy on a regular basis. It has been written about previous wars that only about ten percent of soldiers serving in the armed forces will ever see any kind of combat action, and to encounter the enemy up close just reinforces to me this is no joke.

We took the team back to Song Mao where they exited, giving us a wave of thanks as they moved inside their base for those meager comforts like a shower, hot food, and a mattress, bravest of the brave in my book. The day was now ending as we flew back to LZ Betty, knowing somewhere inside me that maybe, just maybe, we were spared.

THOMAS MCGURN

NVA soldier

22. April Fool's Day 1970, Song Mao

Song Mao, Song Mao, Song Mao, oodles of missions always headed for this tiny village at the base of the mountains. At this period, I had attained the status of a senior pilot in command and only had two long months left till my rotation back home. No rocket scientists needed here to understand the mountains above Song Mao were swarming with both NVA regulars (North Vietnamese Army) and VC local insurgents (Viet Cong). We learned during the night of March 31, 1970 that the fire base at Song Mao was under heavy attack and our flight Platoon would be sending some early morning support to this small artillery outpost.

I have highlighted the below reports which were summaries issued on 30 April 1970, these indicate totals statistics over the period of engagements in the Song Mao area.

On 010300 *(01 indicates the day, 0300 indicates the time BN2643 indicates the grid)* at BN2643, in SONG MAO City, the 2-1 Cav compound rec mortar and rkt fire. Results: Frd—2 US KIA, 13 US WIA; En—5 KIA, 1 SA CIA. On 011410, at BN614476, 7 km NNE of SONG MAO, 3/@/2-1 Cav was engaged v\by the enemy. 1 APC destroyed, 1x M48 tank light damage. Results: Frd—11 US WIA; En—17 KIA, 1 SA CIA.

On 010040 at BN275545, vic SONG MAO, the 44 Regt\. HQ received

100 rds 82mm mortar. Ammo dump destroyed, fuel dump 50% destroyed. On 010030, at BN2643 in SONG MAO City, 2-1 Cav and 44 ARVN Regt (-) engaged the enemy. In the day long contact, the following casualties resulted: Frd—8 KIA (2 IS, 2 ARVN, 4 Civ), 39 WIA (34 US, 5 ARVN); En—151 KIA, 14 PW, 26 SA, 7 CS CIA.

Song Mao being an artillery fortification, the untrained might believe with all that fire power consisting of 105 howitzer M101, 105mm round capable of firing a round 7 miles, mixed with 8-inch M-1 gun 203mm round capable of firing a round 12.5 miles, and 175mm M107 self-propelled gun capable of launching a round up to 25 miles should put the fear of Buddha into the enemy psyche. However, formidable as these weapons are, their usage was strictly long range and coupled with a slow rate of fire and a mass enemy attack in close, made them reasonably ineffective. On this April 1st night, the estimate was that up to three enemy battalions were involved in a coordinated attack at Song Mao. This could represent anywhere from 900 enemy soldiers to 3000 soldiers, a very large force.

How convenient Bruce Britton my roommate and I had been assigned to support units at Song Mao, probably MACV. It was still dark as we both awoke, gathering our personal protection equipment, guns, ammo, survival vest, bullet proof chicken plate (maybe, but who wants to find out), food and water for this long day. Our crews had also been alerted during the night and they would meet us at our respective ships in the revetments at first light. This was one of those many times we would launch before breakfast, fending for ourselves. We all had become very efficient at self-care. I forget who took the lead as we departed northbound for a direct flight to Song Mao. I am often reminded of these irreplaceable moments of visual pleasure when waking early on some future vacations located near any ocean, the calm and peace of such sights. Before the sunrises in the twilight hours, you can see the grey low clouds over the ocean beginning to mix with the first rays of the day creeping forward by the earth's rotation before the sun breaks the horizon plain. As brief as these experiences can be, they give one that moment of total relaxation before the reality arrives.

Almost instantly after takeoff, we directed our vision north toward Song Mao. There was no wind and noticeably visible was a column of grey black smoke ascending a few thousand feet, staining the start of this day's sky and marking the exact spot of the attack on Song Mao. We checked in on the appropriate net (Radio Freq. for supporting Unit) as they instructed us to land on the village pad. The airfield was closed due to the ammunition dump was still on fire and multiple explosions were still occurring, throwing immeasurable shrapnel fragments hundreds of yards in every direction.

Approaching Song Mao, it was clear a major attack had taken place. You could still see explosions from the ammo dump occurring, expended rounds from small arms to artillery shells bursting with hot multiple colors like a man-made volcano. The airfield was covered with exploded and unexploded rounds and the POL refuel area was so badly damaged there were no refuel services available. These guys had suffered a tense night and casualties were high as they instructed us again to land and wait for further instructions.

Short final to Village pad, a clear pad area at the South end of the Village, very close to some shacks. After landing, both ships shut down and after a quick aircraft inspection we did what Army Aviators do best, we relaxed and ate. The sun was now up and you could feel the heat building as we used the ship as a large sun shade while the two crews discussed worldwide issues, example how long will we be here, what are we doing for lunch, and when will they get us flying. A nap is in order. We were aware that soldiers had been seriously wounded and some heroes had lost their lives. I don't mean to expel this tragedy; this was always with us you can't help to think of these young men and their families. It is a constant in Vietnam. Army Pilots would receive a monthly magazine from the Army Aviation Association and the first section we would hasten to review was the Pilot Obits. Sadly, you would search this publication for classmate's names and always find one or more who were felled during their time in this conflict. This practice would continue even after your turn had arrived to depart this war. Always checking to validate that your now ex company pilots, friends, and soldiers had not made this death tracker.

Relaxing, both Bruce and I rested our heads on the aircraft skid which substitutes as a magnesium-aluminum pillow, now stretching out in the shade of the aircraft. Our position was maybe one quarter mile from the airfield which still had detonations occurring indiscriminately from the ammo dump. I can't explain whether I heard the whistle or the contact sound of an expended fifty caliber round as it struck the skid next to Bruce's head (not that it could have penetrated this mindless Neanderthals skull), just missing giving him a monster headache, but continues its ricochet path striking him in the left thigh. This used round did not have the concentrated force as being fired from a barreled weapon, just the speed caused by its overheated casing exploding the projectile from its shell then flying through the air on its mission to wake up Bruce.

After Atlas (Bruce's call sign) calmed down, exaggerating his near-death experience, we all laughed and repeatedly gin up scenarios of what could have happened, imagine if it hit near his crotch. It still would have been a miss.

Missions now arrived to support our troops and our two ships would now be separated for the needs of the unit. As my crew cranked up, we were heading toward the foothills, probably on a resupply mission to troops who had been strategically positioned to block the enemies retreating path. Takeoff and flying was now tracking southwest toward the foothills and the grid of our first stop location.

The grounds near the airfield were like a savannah with grass interrupted by low trees and bushes, however you could see through most of this ground cover below. The crew chief spotted them first and announced this over our internal communication (ICS) to the entire crew that there are bodies everywhere, all over the place.

Once alerted, the eyes went to work and clearly lying there in that motionless battle pose of the dead was the retreating enemy path, laid out in death. These were enemy soldiers who got caught in a shrapnel grinder during the night. This was the result of concentrated artillery support from other fire bases coupled with aviation support from gunships and most likely Spooky, an Air force aircraft armed with mini guns designed specifically to shoot down onto the enemy force. Later, I would be reading that up to two hundred enemy soldiers were killed in this remote battlefield. I would concur after having flown over numerous enemy KIA's over the next two months as their bodies turned black and would eventually melt into the earth, unknown to all in the future.

There would be more missions at Song Mao, and they would challenge every skill I had learned to aid my survival until my time was done in this foreign country, (short timer).

23. Trusted That Guy

NOTHING IS EVER EASY

Just returning from an extended day of flying, I checked in with flight operations to discover our ship would be primary standby this night. I had been flying with Norm as my PC this day, always a pleasure, and my regular crew chief and gunner were assisting in major maintenance of our own ship, so I was now flying a different ship with a different crew. We had just refueled at Song Mao and arrived back at Phan Thiet with just under one thousand pounds of fuel compared to a topped off tank at fourteen hundred pounds.

I was heading to POL to once again top off for the night when my crew chief requested could we go right to the revetment; it would save some time for them to complete their post flight obligations and be able to grab chow. Recognized requests from the crew was always good for morale, while the chief assured me he personally would have the fuel truck top us off.

I always worried about what's ahead, especially when we had been designated stand-by ship, fuel was comfort, one less thing. All the 192nd Avn enlisted crews were some of the hardest working men in our unit, they flew all day, conduct maintenance before and after each flight, and were randomly chosen on many occasions to be reassigned to nightly guard duty, sitting awake in a sand bagged bunker sweating on the perimeter maybe once every five days with the infrequent standby mission always lurking.

Landing, we placed the ship into our revetment, confident the chief would have it topped off with JP4, again, you never can have enough fuel. After checking in with flight operations, Norm and I walked back to our hooch to drop our helmets and equipment. Norm was a little bit upset with stand-by, no relaxing brew this evening since we were on call. But Norm always managed a smile, followed with some satirical remark and made us all laugh with his comical displeasure, enhanced with an actor's face while sucking on a cigarette.

At this time period, Capt. Kenneth Boley of Beaumont TX was the first platoon flight commander with Lt Richard Colbert aka Hoss as second in command. Both were great guys and it was my pleasure to have served with them both. I would again cross paths many years later with now Colonel Boley in his assignment as the commander at the Aviation training facility in Indiantown Gap, Pennsylvania, where I would annually attend my future helicopter simulator training.

As this date's sun sets in the west behind the mountains, silhouetting their sheer size, their green hue turning almost purple as they blended in with the night, they would soon disappear from view in the blackness, like some magic act, but still ominous, like a deadly fly paper. It was not hard to understand all the generations who preceded us for fearing the night.

Normally after returning, we trotted down to our mess hall, if it is still open, for my favorite meal, mystery meat. This again would be some piece of meat swamped in gravy, which we all thought was rabbit. Good gravy makes anything tolerable with the mixed taste of overly preserved bread and sluiced down with varied colored Kool aid, this always served in large metal pitchers. Hunger makes all food suitable. The mess hall was a brief walk north through the enlisted area with the occasional open piss tubes adorning the path, location, location, location. After the meal, either from our mixed stock of C-rations, sent food, or the mess hall, a shower was in order. Hot or cold, it was one of those little pleasures to make you feel clean and refreshed, even if you did walk back in a strong wind with all the red dust trying to adhere to a host. Most of the pilots sitting in front of our building were socializing with tales of what happened that day or what didn't happen with the mix of every topic you could imagine. All were relaxed with legs stretched out from their individual multicolored thrones of nylon.

You could hear the ring of the Platoon leader's crank phone from its olive drab canvass holder. For standby crews, this was a miss a heartbeat moment, that flutter of dread, anticipation, and realization…I'm standby.

Flying at night in the States is not a horrible thing with all the lights and navigation aids, but here in the Republic of Vietnam it was never welcome. Temper this scenario with the varied possible mission complexity and your probabilities of disaster quickly intensify. Captain Boley placed the phone down and advised me to get my crew and report to flight operations for a night Ranger extraction.

Knowledge can be an adversary, having known of at least two failed extractions, one that nearly killed Bruce and one that killed John Wright. But it makes no difference, you have to suck it up and go, it's the job. If not my turn, then someone else's.

Lt Colbert (call sign Hoss, because he looked like Dan Blocker from Bonanza) approached me, requesting to come on this mission as my co-pilot. I told Hoss to go ask Norm if he wants to give up his ride. Now Norm is a Warrant Officer and Hoss is a 1st Lieutenant, but it was still Norm's option, it was his seat, so in typical Norm panache he made his decision and the ever-gracious Norm let Lt Hoss take his place on this very dangerous mission.

Norm was absolutely aware by now how dangerous any night mission can be so this one was not an adverse decision for Norm. He just made his future Platoon Leader pleased, and now can start his ritual nighttime drinking. I felt waggish about Norm's decision, like a karma thing was going on, especially since no other PC had jumped up to ask to take my spot.

Reporting to flight operations as my crew prepped the ship, I was informed that a LRRP Team in the mountains northwest of Song Mao need an immediate extraction. The team believed they were surrounded by NVA in the thick jungle which surrounded them and cannot wait till morning when daylight would aid the enemy encirclement.

These night extractions missions were as serious as you can get, there was no doubt this team was in trouble and could not sit out this endless night, motionlessly huddled together in their jungle cubby until the dawn arrived. Also, at operations was an additional crew from our second flight platoon, they would be the chase ship this night so the respective Pilots in Command worked on frequencies and flight plans. The Gun Flight Platoon call sign "Tiger Sharks" had already launched two ships in case the LRRP team made contact before their pickup.

After leaving flight ops, I went straight to the ship where I saw that we, "the crew", now had an additional passenger, a Senior Staff Sergeant/Ranger who was occupied with installing a wire ladder to the cargo floor, more bad news that this mission was becoming very complicated. Apparently, the PZ (pick up zone) would be in the wooded jungle with the team frozen in their present position. Any attempt at a touch down landing could not be made, a new first for me and the crew.

While at flight school you never received any training in this type of equipment, let alone at night, once again it was on the job training. But who cares, there were brave men in danger and these guys didn't scare easily.

The senior Ranger tied the wire ladder with a spider like mixture of ropes, securing this to several cargo rings on the cargo floor, each hook

CHECK RIDE

could sustain about 1200 pounds of pressure. The sergeant explained that if the aircraft was in trouble, let him know immediately and he would not hesitate to cut one rope which would release the entire ladder. Imagine this, the man was telling me he would cut loose his own troops beneath our helicopter, dumping them through the trees if I ordered him to for the safety of the ship.

This was the scenario that happened to Bruce as a co-pilot, they had a ranger team on a slightly different rig called a McGuire rig when his aircraft started spinning at night and they had to cut the rope on those troops, letting them plunge to the jungle floor, seriously injuring one Ranger, they were lucky. These LRRP teams were the toughest of the tough, so there was never any suggestion that they just wanted an early out from the jungle for a flamboyant thrill.

Once again climbing into the darkness northbound, heading toward Song Mao and those mountains, always the mountains and below the occasional dim light of some small village. Lt. Hoss was a good pilot and I was just as comfortable with him as I was with Norm as we automatically dimmed our interior lights for maximum night vision effectiveness when I now notice **THE FUEL!**

Big surprise and complication, we had not been topped off as I inquired **WTF** happened to the ship being topped off when the crew chief jumped on the intercom to explain the fuel truck was down, as I retorted, "Maybe you should have permitted me to know, chief!"

Spilled milk as I mentally computed the fuel we may need to complete this mission, we still went, the LRRP team could be running out of time.

Approaching our PZ as we check in with the FAC (Forward Air Controller) a Bird Dog Fixed wing who had been in constant communication with the distressed team as we joined the same FM frequency. We were within range to hear the LRRP team's communication.

Unnervingly we listened as they, the LRRP team, spoke in a whisper, the strain in their voice was very perceptible, knowing the enemy was closing on their location, a hint of desperation. Both Hoss and I exchanged turns at the controls to afford both of us the final review of our maps, using our red lens flashlights to check the topography. One last internal brief of the crew and we were ready to do our jobs.

Approaching the hillside, our eyes had adjusted well, thriving in the low ambient light of the ships red and green navigation lights mounted on the side of the aircraft. With these navigation lights on dim, we could start to see vegetation around ten to fifteen feet as we approached the team. The FAC has been calling our location out by clicks as we started slowing down, looking for the LRRP team's blue strobe light penetrating straight up from their jungle PZ.

There was lots of chatter on radios, including our internal now, crew, FAC, LRRP team on their handheld FM radio, still with their attempt at a quite whisper and that hint of fear as we picked up their blue vertical strobe. I was flying, not because I didn't trust Hoss, but I had the experience, something he would glean from sitting in the co-pilot seat on this, his first extraction. This is how we learn, and I was certain I knew a few more tricks at this time in my PC status.

I positioned the aircraft directly over the strobe. The team was directly below as the crew chief and gunner deployed the wire ladder forty feet plus to their location. Looking out the windshield there was nothing but dark, no horizon, and no reference as the crew continuously called in slight movement corrections to aid my control inputs, trying to hold steady overhead these guys. The noise generated by the helicopter could not be ignored. I was sure the enemy in our area had no doubt that we were there and hopefully probably it had them perplexed with this activity.

The rotor downwash was gusting the treetops like some sudden storm ripping past this ridge top. I placed the machines nose with the left side chin bubble (plexi glass window, just below the left pilot's feet)

CHECK RIDE

directly over the top of my chosen tree with one small branch of leaves pushing against the chin bubble. I had fixed the branch and leaves for my reference and coordinated my aircraft control inputs to keep them there like they were glued, a field reference, a fixed position while staring down through my portal. Yes, it's hard not to tense up (I attribute my premature hair turning grey to this mission). This action optimistically insured I would keep the aircraft in one spot, however I had at least three to four LRRP team members attempting to latch onto the hanging wire ladder below.

The team members' actions of securing themselves were now pulling the helicopter down and to the right as our lateral CG (center of gravity) was being maxed out, the sensation was like a large man stepping on the side of a canoe. My small brain was filled with rote movements while attending to the positioning of the helicopter, constant micro corrections of my hands and feet, this was intense, so many things had to go right to prevent that one event from letting us crash, while Lt Hoss checked the gauges for available power and watching our fuel consumption.

My crew chief and gunner were pro's, without their constant verbal adjustments transmitted to me through our ICS (internal comm system) of their respective sector completion of this mission would have been impossible. This including the dual duty of the crew chiefs monitoring of the LRRP teams actions below.

The LRRP team could not climb the ladder into our helicopter. They could only hook themselves to the metal wire rungs with a carabiner and would be lifted straight up until we could clear the PZ with its many protruding branches, like trying to walk through a heavily wooded area without getting slapped in the face, at night.

As our available power moved to its maximum during our attempt at departure, Lt. Hoss constantly updated me. Somewhere in the back of our minds, we knew the unthinkable was always lurking.

I can only imagine being directly under the helicopter, looking up at the dirty side of the bird, taking the team into the night as the crew directed me to hold steady and come straight up. Now straight up is subjective on what I believed was vertical from my position and control inputs. Lifting vertical, your impressions along with your senses, communicate that I'm good but without sufficient other cues, many illusions can be taking place. As the rotor downwash disturbed the tops of the trees, they can produce the perception of actually moving backwards when you may be moving forward or fixed.

This is where the crew chief and gunner earned their keep. Certainly, without their instructions, many aircraft would have been destroyed. Power holding steady, no RPM bleeding as the crew gave me the clear to go; meaning we had attained enough altitude to ensure our LRRP team would not be slapping into the treetops as we gained forward airspeed.

We were out, I could feel the excess weight being transmitted through my appendages onto the flight controls as it pulled the ship to the right while I adjusted the ships position for this critical vertical lift. Flying in nature's blackness, heading down from the mountains, trying to hold about sixty knots (70 MPH), constantly feeling the drag of our LRRP team members below always tugging the aircraft to the right. This LRRP team clenched on with every ounce of their remaining strength for this open-air ride forty feet below the helicopter, swinging in the darkness, thousands of feet vertical from mother earth as we continued our descent from the mountain tops until we could safely position them gently down at the Song Mao airstrip. Balls!!

The cockpit pressure had relaxed a bit as our LRRP team was attached on their Peter Pan ride to safety, but our fuel situation for the duration of this mission to get back home was now becoming critical. Approaching the Song Mao airstrip, we can now use the aircraft's landing light in tandem with our search light to illuminate our placement of the team directly over the safety of the landing area. These lights were not exploited during the pickup phase to avoid the enemy possibly using them as a guide to the detection of ours and the LRRP team's position.

Slowly making our approach while decreasing airspeed, I would hate to come this far and drag these guys through the dirt. Now at a high hover over the strip, once again at maximum power as the crew directed me to lower the team straight down, which we accomplished placing them nimbly on the ground.

I moved the aircraft to the left of the team and brought it down to a soft landing, my opinion. Song Mao was not a secure airstrip at night but the fortification of the artillery compound was within sight and awaiting our boys. However, the Song Mao POL (refuel) facility was closed at night, not helping my fuel dilemma. Our team was safe as they unhooked and approached our aircraft to wave their thanks with their exhausted arms while our crew and the Senior Sergeant disconnected the wire ladder rigging for our express trip home.

Now on final was the other rescue ship who picked up the remaining Rangers off the ridge, relief felt for this LRRP team and our sister ship, tremendous we all made it.

Sitting at idle, trying to save fuel while policing up our equipment, that big red light on top in the middle of the console was now smacking us to attention, MASTER CAUTION illuminated with the panel light 20 MIN FUEL came on. Choices here were dwindling. We had to leave now or choose to shut down, leaving the ship exposed all night to enemy activity. I chose to go, knowing this could be a bad judgment moment. Lifting off, LT Hoss was now flying. I instructed him to straight line our course to LZ Betty and to try and cruise at maximum fuel efficiency. Normally if we had a full fuel load, we would be most likely arriving at base as this warning would flash on, but we started in the hole down 400 pounds. The UH-1H holds approximately 209 U.S. gallons of JP4 and in a two-hour flight we burned off 150 to 160 gallons, so here we were at about 400 lbs. (58 gals.) left. It would be close.

Closing in on LZ Betty, Lt Hoss was a good stick and had flown us directly to the city lights of Phan Thiet, but I had been concentrating on time, We were now over fifteen minutes into our low fuel light, with maybe just under four more minutes to go for touchdown. I was prepared, looking for forced landing areas while running through the steps in my mind to successfully complete an autorotation (**Autorotation** is a state of flight where the main rotor system of a helicopter or similar aircraft turns by the action of air moving up through the rotor, rather than engine power driving the rotor) at night because I was sure our engine would quit.

I was in charge of four other souls on board here, GOD let us make it. Those last few miles were the longest in my life, now twenty-two years old and not smarter after all. I wondered if I'd be there for twenty-three as we touched down, rushing the shutdown checklist before lack of fuel did it for us.

The Senior Ranger departed after handshakes, walking with that satisfaction that one of his six-man teams had made it out of danger.

Our crew gathered for a cursory debrief, Lt Hoss was stoked. He wanted to write us all up for some kind of citation, but I told him we didn't do that much in 1st platoon. The crew chief was apologetic about the fuel circumstances, but his actions were exemplary, and we made it home. I complemented them both as I said without them, none of us would have come home on this one.

Lt Hoss accompanied me to flight operations and after checking in we took that red dirt path back to the officer's hooch where most of our pilots have long since hit the rack in preparation for their own dawn patrols.

Next morning, Norm woke me up with the sun way past the horizon. We had a late mission and he had stepped up by completing all the PC tasks. Norm questioned me about the extract mission and I told him, you missed a good one. Norm was a good friend and it was time to cut him loose and move him up to PC.

Looking back, I would be keen on knowing what happened to this LRRP team of great Americans. I hope they all lived a full, contented, and Vietnam dream free existence. We had landed with approximately seven gallons of fuel left in our tanks, and most of that would have been unusable. Now after some forty plus years, I will always fill my car before long trips and vowed to never run out of gas, (Another Brain stamp).

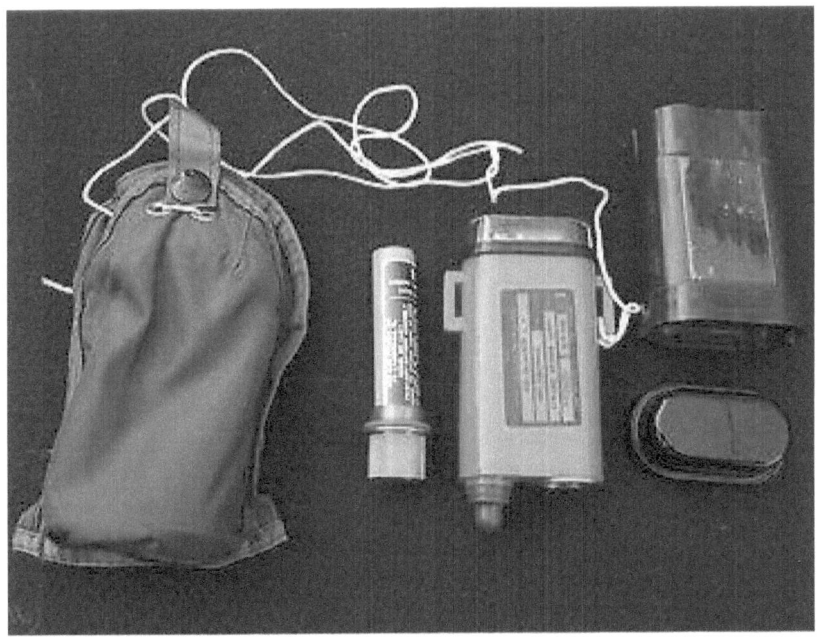

SDU5 strobe light kit

VECTION ILLUSION

This is when the brain perceives peripheral motion, without sufficient other cues, as applying to itself. Consider the example of being in a car in lanes of traffic, when cars in the adjacent lane start creeping slowly forward. This can produce the perception that you're actually moving backwards, particularly if the wheels of the other cars are not visible. A similar illusion can happen while taxiing an aircraft.

Wire Ladder

Ladder extended maybe fifty feet.

24. I Thought They Were Gone

Just one of those everyday common missions supporting some mechanized unit northwest of Phan Thiet on the hunt in the brush forest west of Highway One. These mechanized units were composed of APC's usually armed with the M2, 50 cal. Machine gun (Armored Personnel Carriers) used to safely transport a patrol team to locations while sitting inside their protective shell from small arms fire.

They often were accompanied with the M60 Patton Main Battle tanks. Working together, they would sweep into the thick brush forest of the Le Hong Phong which covered both sides of Highway One, making a perfect blind for enemy ambushes. These tracked war wagons would penetrate through this maze of twisted trees and vines, searching for enemy locations or storage depositories to deter future attacks which were common along this thoroughfare.

This mission was a dedicated mission which meant you and the crew had been assigned to a specific unit for the day to assist them with any of their priority necessities of that day. Such precedence could start out as logistical supply pickups to disperse through their command which we were supporting, also always the movement of personnel. Soldiers who may have served their time would receive that well deserved flight to Cam Ranh Bay for their rotation home, boarding a freedom flight. Or just a command and control flight directed by the unit commander to view the positions and operation of his individual squads.

Having flown a number of such missions, it was always interesting to view the positions of these mechanized units. At night, they would circle their machines like wagon trains on some American prairie to cover all axis of possible enemy attacks. I wonder how far back in history the circle defense must go, unambiguously proven through time. While in this formation, these units nightly would encircle themselves with concertina wire or Dannert Wire which is a type of barbed wire/**razor wire** that is formed in large coils which can be expanded like a concertina. Just another layer of defense to inhibit Charlie, but our enemy could get so close.

During daytime operations, these units would first have to break down their perimeter by loading this makeshift defense equipment, making it compact enough to be carried and stored on their machines before they moved out in search of enemy pockets buried deep in the maze. A typical day started with breakfast, then breakdown perimeter defenses, move all day through unbelievable heat, dirt, dust, and thorny brush looking for the enemy, then to establish a new camp, deploy perimeter defenses, have dinner, and finally have a peaceful night surrounded by possible enemy sappers (defined as military engineers but in Vietnam they were those sneaky bastards that wormed their way through the wire defenses). Once again, I had chosen the right path of Aviation and was reminded of this constantly with such scenes of misery.

By late afternoon, most of the mission had been accomplished and we were now flying the unit commander as he conducted an over watch as his APC unit pushed through heavy brush in search of signs of enemy activity. Slowly the track vehicles moved snake-like, weaving through this wooded tangle, kicking up a cloud of dust for those who followed to ingest. The APC's pathway was clearly visible as it was the dry season as they crept forward when it happens!

We were flying overhead in a left-hand pattern, positioned directly over the lead APC. The unit commander was conducting his instructions through FM radio communications to the troops below. I was looking vertically down at the lead APC when a blast directly under the lead APC, with that immediate obscuration of dust and expelled explosive materials totally engulfing the machine with the concussive force reaching skyward almost immediately.

My first impression was, they are gone, all watching believed we had just witnessed the multiple loss of American lives in an instant of war. The column came to a halt and radio chatter quickly commenced again after that stunned pause of the unbelievable was assimilated. Smoke and battle dust began to clear. Amazingly, the APC seemed intact, a miracle, but those inside must have been scrambled like frail eggs. Quickly, the lead APC was on the radio. They had injuries, but minor injuries. One soldier's foot had a severe cut when the blast almost penetrated the APC floor, cracking the armor. The rest of the crew were stunned but functional as they started to receive assistance from the following caravan of comrades.

Charlie was smart and, on this occasion, they had found an unexploded U.S. 105mm artillery round and booby trapped it (In the late 17th Century, hungry sailors would set a trap for a seabird known as a "booby." The term "booby trap" was literally "a trap for a booby." However, it has evolved to mean a harmful device designed to be triggered by its unsuspecting victim) now strategically placing this device in the best possible position to be lethal against their enemy, us. This bomb could have been hastily placed by the enemy minutes before the APC encountered the device or triggered manually as they watched from some hidden observation location. Even Charlie makes mistakes and their miscalculation resulted in our soldiers' lives being spared. They buried this device too deep, diminishing its useful explosion and saving the APC from catastrophic destruction.

Between NVA and VC, it was estimated during the duration of the Vietnam War that their death casualties were close to two million. So, in this diminutive country you can understand that they, our enemy, seemed to be everywhere and still managed to keep their commitment with such losses. Imagine the force that wasn't killed. Mission over and we once again returned home, appreciative to be flyers.

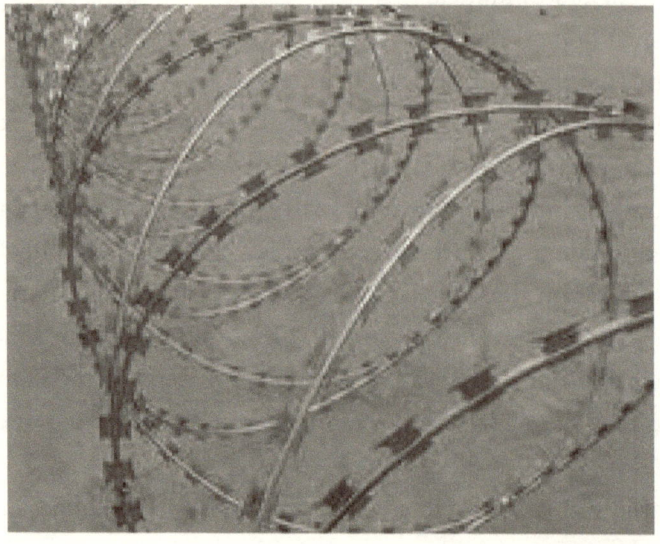

Concertina wire (Concertina is like an accordion) this razor wire can be compressed.

CHECK RIDE

*The **M113** is a fully tracked armored personnel carrier, commonly called an APC.*

APC with 50 Caliber Machine Gun Mounted on front

25. The Real Times

Flying missions, every day lots of resupply to various units in the bush, long days with tight LZ's, the unexpected always lurking and occasionally seen. One such mission we were in route to the toilet bowl jungle area which I have explained in previous stories, not a nice place. This was one of those sad days, although the sky was clear and flight conditions were perfect.

This mission would remind oneself of what could be. Normally all crew members had body armor in hostile areas, but it is heavy, hot, and restricting, so we only wore it on the perceived dangerous runs. Today was one of those. The pilots and crew wear a chicken plate which covers us from our waist to the neck area. It is composed of a Kevlar fabric, next is a ceramic filling, and finally an aluminum metal wafer material, usually worn over or under our individual survivor vest which was stuffed with all the trinkets needed for survival. This armor was supposed to stop a 7.62 caliber rifle round. As additional protection, the pilot's seats were also armor plated, but we all know one well-placed round or lucky shot could bring down the whole show, perception versus reality.

We had been informed to proceed to a unit's grid location and drop ammo, water, food, mail, and on the return make an honor escort of a soldier who had been tragically killed in battle.

Customarily, Medevac would pick up our mortally wounded soldiers but as I have related before, these ships seemed to always be overworked…a genuinely tough mission day after day.

These jungle LZ's (landing zones) can be very rough. Many of them had been blown out from the jungle trees by the troops wrapping each tree base several times with detonating cord, a kind of plastic explosive tubing filled with C4 plastic explosive. This practice left numerous stumps which look like giant jagged toothbrushes. It's not like raking leaves at home this is exhausting work as the ground grunts are always multi-tasked, now you know why they are called grunts.

These areas are always small and tight, not fun, often called hell holes for the life-threatening challenges they present to helicopters. Crew coordination was so essential and the only reason we could get down in such areas safely, it was not parallel parking. Constant intercom directions consisting of up, down, left, right, tail left or right and elevation above ground, teamwork, teamwork, teamwork. Thank God for the crew chiefs and gunners. They saved the Army billions worth of undamaged aircraft.

We did our best for the soldiers in the field. We wanted them to recognize that we were there for them, that is the whole point. Politics, battles, results are not at this level. It is always about your fellow soldiers.

Landing now, dust and vegetation was being sucked up and down through our rotor down wash. The landing area was not level, stumps were everywhere as the crew talked us down to that wobbly ground contact. As you looked out at this small perimeter, finally sitting static with rotors at flat pitch, now seeing these soldiers watching us, and probably wishing they were getting onboard to leave their misery here in the new no-man's land.

Other worker bees hurried to empty the ship as all the supplies were quickly thrown to the ground, plastic gallon bottles of water were one of the most appreciated deliveries. We could now see several soldiers carrying their expired comrade, respectfully hand moving his remains to our ship. Every soldier in that small blown out hole in this jungle was now focused on their fellow soldier, brother, friend, who was wrapped in a poncho liner with only his legs exposed. You can sense the dead weight even though the soldier's body appeared to be a slight fellow.

This was a field funeral now taking place before us, a solemn procession to this soldier's aerial hearse, and this will be the last time his buddies shall see him. But I'm sure they will never forget him or his unceremonious departure from his last stand.

I think I remember this soldier was a Lieutenant. His boots were visibly worn, needing polish, one of his dog tags had been laced through his boot, a practice during this war brought on by the necessity of identification from explosive impacts which caused some very devastating bodily injuries.

This would be this soldier's last flight on an Army helicopter to start his journey home to those who loved him. All we knew was he was killed in action during a small fire fight which occurred within the last two hours and now he was onboard with us.

We took off, the helicopter doors were open with this hero's corpse secured and lying on the cargo floor, the start of his path home. Most likely when his unit will have returned to base from the field a memorial ceremony would take place with a rifle and its bayonet stuck in the ground next to a pair of empty boots and a helmet as his commander speaks words of remembrance, but this noble young man's remains most likely will have probably already arrived home in the States by this time.

These missions always made you retool your thinking about your time in country and we felt sad that this was the best we could offer this young soldier. As we flew his remains to LZ Betty, I for one would say a prayer for him and his family, maybe this didn't accomplish anything, but it made me feel better. Citizens back home watched the nightly news and would scan the papers, whichever way they slanted stories, maybe engaging this day by talking about the war, but this is as close as you can get to understand the consequences and pain being inflicted by this war or any war. And by this, I truly mean both sides, everyone has family. Distance makes such stories of one soldier's death not newsworthy until years later when total deaths are displayed to the horror of a shocked public (statistics 58,800 Dead). War is the politicians' failure.

Mission day is now over the ship is put to bed, maintenance completed, some food, some drink with laughs as all the pilots discuss their flight day good or bad then a final look at tomorrow's mission board and hopefully a full night sleep, our life continues.

I pulled this picture form the internet it speaks for itself (not my mission)

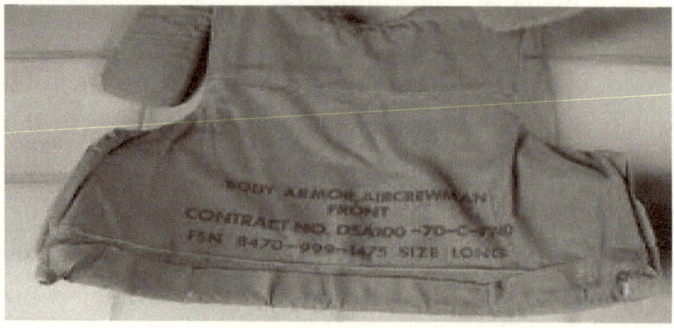

Chicken Plates

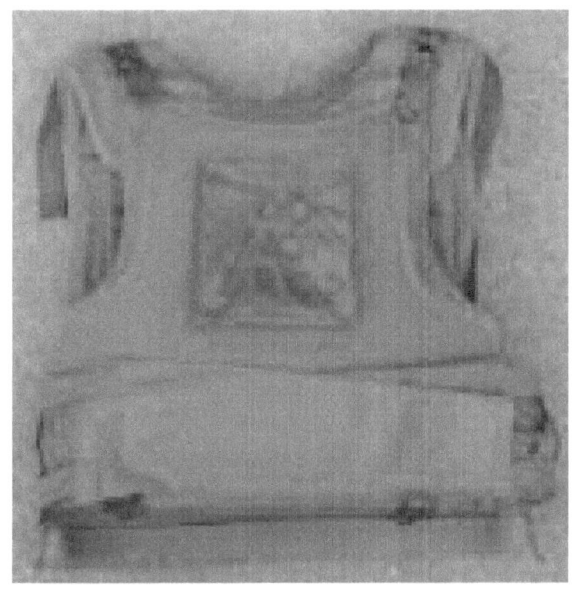

Pilots "Chicken Plate" Ballistic Body Armor front

26. Gotta Go

We would be supporting units in the Song Mao area for tomorrow's mission which had been posted on the assignment board, scribbled in grease pen the night before. The mission board information usually arrived late as mission assignments slowly filtered down to their final scheduling by our Platoon Commander, currently Captain Boley from flight operations. Normally, 1st Platoon pilots would all be sitting around bull shitting about everything and anything, no rules on offensive comments. It was a predictable night, hot, dry, and gritty with the red sand gently blowing everywhere, no grass here, while all of us were waiting patiently for our next day's flight assignment. I would often like to take a shower before hitting the rack, it cooled me down and I felt fresh afterward, making it much easier to slip into that deep sleep, no air conditioning, just tropical breezes from the ocean below the cliffs.

The shower was located just north of our hooch, co-located with the latrine and a large water tower mounted on a sturdy wooden frame which was used to provide water pressure, it resembled those water towers you see in western movies near the train stations. The source of our shower water was tanked in from the Ca Ty River, which ran through the city of Phan Thiet. It was definitely not treated, not potable, and sometimes smelled of diesel fuel. On some nights, you could see attacks taking place out in the flats about eight miles to the west, usually directed at two separated fire bases, one called Sandy. We would then climb the wooden ladder to the top of this tower, extending our view of the action during such an attack. Often you could see red tracers and green tracers flying wildly, while large impacts were seen with a flash and a late report as the impact sounds took some time to arrive, like the thunder of some distant storm, macabre entertainment.

Eventually, after some period of time, a spooky gun ship (Air Force Gun Ship) arrived, raining down a snake like red ribbon of fire from their multiple miniguns as our eyes tracked these tracers onto suspected enemy locations. This vision was then followed by the muffled but continuous angry roar of the miniguns with that distinctive ARRRRRRH sound. Deep inside we yearned for our comrades to be spared, but we were happy to be someplace else. The enemy was impartial, always attacking one base or another, including LZ Betty with one attack being a significant penetration in May 1970. Now after such a diversion, it was off to slumber, hoping no mortar attacks or standby alert horn sounded to disturb someone's crew rest.

Up early, maybe five AM, emerging daylight but still with that gray variety of light that just wanted you to move unhurried. It was either early April or mid-April. My life chapter in Vietnam was shortening rapidly. I felt I had reached a peak in my flying skills, confident now of any mission assigned, seasoned. However, we all suffered the drain of this pace with weight loss, sickness, and fatigue. When I first arrived in country, I weighed 220 pounds at 6'3" inches. When I left, my weight was 170 pounds. But the truth is, one doesn't feel the transformation until you collapse into your seat on the freedom bird flying home.

Skipping breakfast that morning, I would feed myself later with C-Rations or LRRP rations, a dried packaged food which you just added water to, and most likely washed down with a Pepsi or Coca-Cola. We had little confidence in the water here. I had already had a case of dysentery and food poisoning. Soda was a treat and a safe beverage we could easily stock up on, always welcome to all the crews. The platoon's supply was supported when any of us were able to fly to huge bases like Cam Ranh or Phan Rang where the Air Force always had a substantial PX. We would purchase as much as the ship could hold; I mean cases stocked to the ceiling of the cargo compartment. It was one liquid which wouldn't get you sick.

Arriving at the ship, smiles and laughs with the crew. We were prepared for the day, whatever it may deliver, a confidence/bond molded by constant missions, a pleasant work environment.

Take off was maybe seven thirty as the fragrances of the city and ocean once again combined as we sped over the northern boundary of LZ Betty, the perimeter dropping low just after the runway threshold, bearing directly to the mouth of the Ca Ty River. We continued low to enjoy the wonders, hundreds of multi-colored sampans in the harbor with people everywhere on the soiled city streets, Asian life, still exotic, a reminder we were nonetheless the outsiders.

Flying low level north on the beach, with the palm trees hanging just over the edge of the sand, letting the copilot enjoy the controls, remembering my elation of getting to handle a machine back in the day. We eventually started our climb to fifteen hundred feet, the safe altitude standard and turned on course to Song Mao for this day's support mission. These were the moments for most Army Aviation crews where you could appreciate how fortunate one is to be a flight crew member, now flying halfway around the world, doing things the majority of the populace will never experience, but there is always a price.

On short final, we checked in and were immediately instructed to go to the logistics pad alongside of the rugged runway to onload for supply runs. The ground crew had stacked several separate piles, mostly consisting of C-ration cartons, ammunition crates, and dozens of plastic gallon bottles filled with potable water. From experience, I would say water was the most demanded essential, even more than ammunition.

Quickly on this topic, I was flying logistic supplies during the dry season to an ARVN Unit (Army Republic of Vietnam) who were being mentored by two American officers in the bush at the base of the mountains. We had just landed with their supplies, mostly water, and just after our departure the US officers called on their FM radio to report that the ARVNs would not share their water supply. You could tell they were distressed.

Shocked and pissed to receive such a desperate message that our guys were mistreated, I immediately instructed these two officers that we are now turning to return immediately to their LZ, and for both to be prepared to board, explaining that it would be our honor to just take them to LZ Betty for water. Having never met these officers, it didn't matter, they were our brothers in need. Amazingly, the ARVNs quickly shared their H2O supply and we received a very grateful thanks from these American advisors. How do they find men to do this? Soldiering at its best.

Several logistic runs were completed, intermingled with two POL refuel stops. We now had returned to Song Mao airfield, shutting down and pausing until the unit we were supporting organized their next mission priority for this day. Our ship at rest now looked like it has been disassembled. All the doors were open, the pilots' doors were opened forward like a car, some inspection panels were open (it never hurts to give the machine an additional preflight, who knows what wiggles loose).

Lunch was in order, maybe a can of turkey or a spaghetti LRRP ration, and beverage of warm water or a warm Pepsi. The crew chief and gunner requested to walk over to the town at the north end of the field, booty call maybe, permission granted, no urgency perceived here. These were our breaks, our down time. It was never 365 days of war, maybe a quick cat nap.

A jeep was now coming down the runway. It appeared the unit had a new task for us, the day resumed. Two soldiers quickly approached with an urgent mission and the request was totally unexpected. A ranger team in the mountains above had been in contact and they needed an immediate extraction. The soldiers gave us the team's coordinates, which we urgently copied and posted on our map, along with the contact frequencies. Our chief and gunner had not returned, so I instructed the jeep driver to locate them at the north end of the airfield.

Myself and the copilot buttoned up the ship and climbed in, starting the run up as the chief and gunner returned, rushing to their stations, their brief would be over the intercom. No chase ship on this one, no guns, this team wanted out now, and we were the closest aviation asset that could complete that task.

Dialing in the appropriate frequencies, we departed the airfield, turning northwest, climbing and pulling in the collective as we received that feeling of unbelievable power of the machine. Focused in the midafternoon light, we were reading our maps to recognize the contours ahead to facilitate a direct path to these warriors.

The country of South Vietnam was approximately 68 thousand square miles, slightly larger than New York State with their 59 thousand square miles. The mountains in Vietnam seemed to be saturated with NVA and VC using the fringe area of these hills to prepare and launch multiple attacks. By 1970, they had years of preparation. Consequently, undesirable enemy encounters were inevitable. All we knew was the team had made contact with some enemy element. The size and composition were secondary to the fact this team was on the move to escape a larger escalation where they could be overwhelmed.

Time was the LRRP teams' antagonist here as they hustled to their pick-up zone through jungle vegetation, now obstacles to their quick movement. We contacted the FAC aircraft who initially relayed the teams' urgent request and now confirmed our position while climbing up the steep valleys toward the PZ. Hopefully we would be in range of the team's FM radio, making direct contact to guide us in for an expedited pick up.

Normally, the all too distinctive beat of the UH-1 helicopter might be a deterrent to enemy ground troops who were always concealing themselves to avoid detection. It always brought massive fire power against them; they hated us, the manned drones of 1970. One thing about the American Army, expenditures are not a factor, whether ammunition or monetary, when trying to hunt and kill the enemy, or to support our troops. We were all in. But now with the LRRP team being in contact, our unique signature sound was an alarm to every NVA in earshot that guided them to our team. The enemy also had communications, using their resource effectively to trap our men, time could be running out for them.

We were in the area and made contact as the team leader gave us a brief on their PZ. His voice was stressed, or exhausted, from moving as rapidly as he could toward a small clearing on a stream bank as we told them to pop smoke. It was not always the best practice to initiate with LRRP teams, but this was possibly a onetime shot for this team as they popped a green smoke, which we quickly observed and confirmed. Fast tracking this extraction our ship is on the way in!

We were on final heading to the bottom of a small valley with an opening near the stream bank. The smoke grenade was still wafting in the PZ, you could see a corner of the stream where there was a clear bank to land. The area was surrounded by jungle grass on all sides, the really high stuff, hell you could hide a tank in that PZ.

The crew chief, Mooney, and the gunner, Dudley, had their M60 machine guns up and were ready, fingers on the trigger. We were almost there, and still couldn't see the team. I felt a very anxious emotional wave as we touched down, hoping they didn't have company.

The skids sank into the mud and grass bank with the jungle grass right outside out pilots' door window as our rotor wash blew the tops of this overgrown carpet every which way. Then suddenly, like in some cheap horror movie, a painted unknown face is at my pilot's door window, only separated by the thin plexiglass; my heart missed a beat, maybe two.

I was on the controls as that instant surge of holy shit ran through the mind, too slow to first recognize this painted animal as a LRRP team member giving me a happy thumbs up. If this was the Olympics of combat boarding, this team won the gold as they all leaped into their aerial lifeboat. The PZ instantly erupted in machine gun fire, but only from us as we climbed over the jungle wall of trees, escaping whatever was stupid enough to follow this team. I don't think the shooting stopped until a good twenty seconds went by, an exceptionally long time.

Airborne and turning down from the mountains, the team continued in their exuberance, that type where you're so scared it's the end, to that when it was unbelievable you fucking made it. Maybe they, the enemy, were closer than we thought.

My copilot was flying toward the airfield at Song Mao as the team leader positioned himself directly between the pilot's seats. He was a very black sergeant with the most stunning set of white teeth revealing his infectious smile. There was no mistaking this guy, he was America's best. He smelled of jungle, mixed with exertion and sweat, and was pounding me, the PC, on my right shoulder, loudly shouting with great happiness, "Yeah, Yeah, Yeah!"

How can I describe my state of mind that we just may have assisted in saving this team from a horrible fate? CHARLIE would have killed them all, never to be accounted for, just some future bracelet with their names on some American wrists back home. Hence, so exultant is this sergeant that he keeps engaging me to look back over my right shoulder when I notice the bright red slow flow of blood from his forehead, straight down, dripping across his left eyelid and landing on his cheek bone.

This man is wounded, was initially my first impression, but on closer examination, my eyes are drawn toward the source of the blood flow, which originates at his hairline and then I detected that the hair was silky black and very straight. Holy shit! It's a scalp, an enemy scalp, a very fresh one. The sergeant had caught my startled attention as he moved back to hugging his men while my copilot and I thought just how close was the enemy and what would be their response if we were shot down and captured? My best guess was more wrist bracelets with our names, just thoughts.

If you look at the history of Americans captured in South Vietnam, they were minimal, and these LRRP teams would have been tortured and killed. There were over five thousand helicopters lost in the Vietnam War, many survived their crashes and were rescued, but in some crashes, many of our crews were never accounted for. In the south, it was a merciless land, and we knew this, the enemy had no time or resources to care for prisoners.

Landing at Song Mao, the Rangers unloaded but all shook hands or banged on our doors to give us our reward and their sincere thanks as they walked away with that worn-down shuffle. We finished that day's mission with one of those more involved accomplishments we certainly can cherish for life.

Some will say reading this remembrance that war atrocities, killers, animals, are not the American way. I was surprised while researching a Ranger web site to read where Ranger teams were being challenged by their commanders on the legitimacy of their contacts while requesting immediate extraction and would actually bring back anything to prove to their commands that, "Hey, asshole, I'm not lying, we needed to get out of there."

By 1970, every soldier knew the war was a political mess being previously directed by Washington idiots (whiz kids) like McNamara who paid little attention to the Joint Chiefs of Staffs and many years later admitted he knew the war was a stalemate. He was a true political slave, too petrified for his own future to speak the timely truth which may have spared thousands, the definition of coward.

What employee has never been confronted by a skeptical boss? I have listed below a passage which enforces that LRRPs, if captured, would be mutilated. For those who have never had to face such dangers, speak carefully and not for political purpose for these men sacrificed for our Country, directed by politics.

In 1965, in response to stepped up military activity by the Viet Cong in South Vietnam and their North Vietnamese allies, the U.S. began bombing North Vietnam, deployed large military forces and entered into combat in South Vietnam. McNamara's plan, supported by requests from top U.S. military commanders in Vietnam, led to the commitment of 485,000 troops by the end of 1967 and almost 535,000 by June 30, 1968. The casualty lists mounted as the number of troops and the intensity of fighting escalated. McNamara put in place a statistical strategy for victory in Vietnam. He concluded that there were a limited number of Viet Cong fighters in Vietnam and that a war of attrition would destroy them. He applied metrics (body counts) to determine how close to success his plan was.

President Johnson and Secretary of Defense McNamara

Although he was a prime architect of the Vietnam War and repeatedly overruled the JCS on strategic matters, McNamara gradually became skeptical about whether the war could be won by deploying more troops to South Vietnam and intensifying the bombing of North Vietnam, a claim he would publish in a book years later. He also stated later that his support of the Vietnam War was given out of loyalty to administration policy (POLITICAL COWARD). He traveled to Vietnam many times to study the situation firsthand and became increasingly reluctant to approve the large force increments requested by the military commanders.

McNamara said that the Domino Theory was the main reason for entering the Vietnam War. In the same interview he stated, "Kennedy hadn't said before he died whether, faced with the loss of Vietnam, he would [completely] withdraw; but I believe today that had he faced that choice, he would have withdrawn. (Passage research from the web)

Jungle grass

With painted faces, radios, and lightweight gear, the patrol carried heavy ammunition of magazines, frags, smoke grenades, claymores, and often weapons of the enemy, since the M-16 rifle had a distinguished signature. Everyone performed duties, including the team leader, assistant team leader, Kit Carson, medic, radio telephone operator (RTO), and point man. A security wheel of members with one staying awake at all times would be formed at the monitoring site off enemy trails or underground tunnel homes. Claymore mines were spread in front—hopefully in the direction of the enemy. Few sensing gadgets were present and everything was examined personally. The numbers and style of tire shoe marks were noted. Morale of the enemy was sensed, along with their weapons and luggage.

The three to four-day mission did not permit talking, snoring, noise, smoking, or excreting. Urination was permitted by twinkling down twigs to avoid noise. Coughing was not allowed, a muffled cough could alert the enemy. Often the enemy would be within ten feet of the team. The quiet allowed the senses to notice so much: the sudden snap of bamboo growing pains, a jet-like whines of mosquitoes, dive-bombing flies, and butterflies alighting on the guns; the darkness so black that the only visible light was the luminous glow of decaying leaves. Radio contact was frequently by code clicks rather than voice. The food was dry LRRP rations with the water carried. Most often the Rangers were not hungry and waited to eat when they returned to basecamp. There were long hours of tense waiting in the jungle, with feelings of doubt and fear, Rangers coped with many anxieties, including the *possibility of mutilation* by the enemy. Passage research from the web, Ranger pages.

27. It Happened So Fast

Monday February 23, 1970 and I had been a PC since December. I was now twenty-two years old, yea! I only continue to emphasize this fact now, looking back forty plus years, realizing the whole spectrum of luck/karma which melded together to get me home. Will this year ever pass? On this day's mission, we were supporting an Infantry commander's reconnaissance in the mountains of the Central Highlands northwest of Phan Thiet. This was a common practice and usually conducted in preparation for some pending insertions in the area.

LZ Betty had transformed since my arrival, the enlisted areas for the 192nd all now had the same wooden constructed billeting with tin roofs similar to the officers' quarters. The prime exception was the officers only had two men assigned to a room, not the ten men the enlisted had, with all cramped into a sweaty hut little better than a concentration camp. Enlisted soldier's life here was always a struggle. One item all the pilots had was a Julian calendar to torturously compress time as we counted down our remaining final days, one after another, like a child waiting for Christmas. I would sometimes return after a mission to make some special notation of why I should remember that date, and I have this same calendar framed, hanging in my home these many years later, reminding me of those times that need not be repeated.

It was late morning as we proceeded toward our target area in those rugged mountains, how fortunate we were that Phant Thiet had claim to the best weather in all of South Vietnam, just one less thing to worry about. Our reconnaissance pacs now onboard comprised three officers and one senior enlisted man. They had requested this mission to get an up-close snapshot of the terrain and conditions they would be plunged into tomorrow during their company's scheduled combat insertion. Planning in advance for these types of insertions varied but higher command perceptively had established a temporary field artillery fire base on top of an elevated plateau at one end of the nearby valley, moving the light airborne 105 howitzer artillery pieces into position with CH-47 heavy lift helicopters. That gave the ground troops the security arch of the howitzer fire sorely indispensable to our boys who were always at risk of being enmeshed in a jungle engagement.

One of those military phrases I have picked up is steel rain which refers to the numerous steel fragments generated when an artillery round is timed to be an airburst explosion, causing hundreds of steel splinters which cover the area below, devastating to our enemy. Often on many occasions while landing at many diverse points during my tour, you could step out of the machine and find such remnants of an artillery burst with some pieces the size of one's hand. Jagged chunks of steel with razor sharp edges, a terrible weapon used against enemy infantry and a future archeology clue of how long this conflict continued to disperse this amount of combat residue.

It was a beautiful day to fly, the sky was clear and the heat level tolerable before the sun had reached its apogee as we arrived and began our orbits above several potential insertion LZs. The terrain below was exceptionally steep, covered with triple canopy jungle, the thickest vegetation imaginable, with minimal/few clear landing areas accessible, none larger than a two ship LZ at best. My copilot was flying, receiving directions from me and the ground commander as we flew left hand orbits, staring intently below scanning an area where our boys will be deposited. The commander and I both had our maps out, plotting these possible landing locations and mentally calculating the logistics of delivering a company of troops into this environmental puzzle below; lives depend on such decisions.

We flew back and forth and zeroed in on two possible openings below, one was a flourishing green oasis surrounded by lofty jungle trees but only large enough for two ships with a skilled crew to land. Visualize this ground commander's dilemma, here is a captain or major, maybe late twenty's or early thirties, knowing that being dealt this handicap of only a two ship LZ, just twelve of his men/soldiers would be placed on the jungle floor, exposed for several minutes, while having to secure an unknown area surrounded by dense jungle bush holding back God knows what threats before the additional flights of reinforcements' could touch down. Not Normandy Beach, but proportionally it could still be a catastrophe. The ground commander was satisfied as we flew home to LZ Betty where he and his team would complete their detailed planning for the remainder of his day until tomorrow's lift mission. We dropped this recon team on the logistics pad and headed to hot refuel to continue our day, but joyfully received a message from flight operations to shut down and standby at quarters. Joy!

Many times, crews would complete their missions and receive such messages of fortuitous unexpected down time, immediately followed by landing, then completing the shutdown of our ship after it was gingerly positioned into its the protective revetment. At that point, the crew slowly proceeded to their hooch, usually carrying all our "accoutrement" including chicken plate, helmet, survival vest, and personal weapons, and before a nap maybe a binge raid on your personal supply of canned food while sitting comfortably in your beach chair scanning over the dirt courtyard.

Often during such times, you would just sit and watch the gun platoon area, kind of like the monkey house at a zoo, unpredictable. If they, the gunnies, were not out on a mission they would be sitting outside their rooms often cooking, joking, or smoking. Their lives were more leisurely, only flying when necessary, but their exposure to enemy fire during such missions was substantially higher and no one felt jealous of their down time, they were different. Like most large Pods of men, they will breed their own environment of games/practical jokes or worse, but in a war zone it wasn't just lighting a bag of crap in front of someone's door, it was amplified.

There were at least three occasions that the following occurred. Phan Thiet was a City on the Sea and at night the Ocean breeze would bring the pungent smell of nouc mam (**Fish sauce** is an amber-colored liquid extracted from the fermentation of fish with sea salt. It is used as a condiment in various cuisines. Fish sauce is a staple ingredient in numerous cultures in Southeast Asia and the coastal regions of East Asia,) not pleasing at all wafting from the north to the south.

The enlisted crews were sheltered to our north, and knowing which way the wind blows, they had perfected the use of wind power by setting off a large CS container of gas, not the small grenade type but the big canister ones which we deployed from our helicopters with a timing lanyard and they did this late in the evening, thus sending this white milky cloud of temporary pain directly through the officer compound, sparing none in its path. (CS gas is generally accepted as being non-lethal).

The alarm quickly sounded, gas, gas, gas as we all scrambled, driven with the memories formed during our basic training of that dreaded gas chamber experience while now climbing over each other, looking like fire had been applied to an ant hill, breaking out our aviation gas masks, trying desperately to avoid the burning tears and running mucus, but knowing we had been out-maneuvered by our enlisted crews.

Oh, how they must have been in their joviality that they had interrupted our opulent lifestyle. However, that action has disturbed the beast.

On this one occasion, one of the gunnies known to be a little intolerant, promptly charged from his hooch like a bull blinded by rage, sprinting to the north edge of the officer perimeter embracing his carbine like some mad movie actor in a last stand scene, then immediately discharging a full clip of live ammunition in the direction of the enlisted area, although his elevation was substantial enough to deliberately ensure there was no collateral damage. This heroic action was supported by some alcoholic beverage, no doubt. Like I have already stated, they, the gunnies, were different, and this officer's solution secured that this was the last time we were ever gassed by friendlies, well maybe not so friendlies.

Nuoc mam Factory

Our break that day turned into an extended down period, very, very welcome, early shower, meal at mess hall, and lots of personal time. Later that evening, the mission board was posted, and I had been assigned to tomorrow's lift. The mission briefing would be after breakfast, around 7AM.

Up early and walking out our screen door, the courtyard was quiet. My roommate Bruce had also been assigned to the lift mission, but he was from the West and his internal clock was always three hours later than mine. Bruce continued to sleep, every nerve in his body shut down, like an old dog. I always loved this time of day, gathering my wash kit to leisurely walk down the wooded planked path to the wash sinks to shave and brush my teeth, all conveniently located next to the latrine/outhouse with its own unique aroma of old wastewater and soap. With all these combined scents slightly overpowered by the five in a row wood seat latrine with the cut down steel collection barrels underneath.

It was another perfect day gazing out to the South China Sea as the low clouds now received their golden edges from the sun while the fishing fleet was already in place, dotting the ocean, all bobbing to the ocean sway. What a gift to enjoy these visual moments, an awareness interruption.

Back at the hooch I was soon climbing into my Army shorts and tee shirt, followed by donning my custom made two-piece nomex flight suit, ready to rock, off to the office. Breakfast first. After a short walk through the enlisted hooch's area, I turned the corner to enter the Mess Hall, now serving the early birds, no need to look at the menu, it was the usual: eggs, bacon, toast, coffee, tea, maybe grits with juice.

Tuesday February 24, 1970...the briefing room was located in flight operations. PCs and most copilots were gathered for the day's mission scenario. Thomason was again flying with me. He was also my copilot during the previous day's reconnaissance, a really capable guy who would be moving up to PC in the next couple of weeks. The Commander of the ground troops gave us the mission overbrief, detailed with maps and grids of insertion locations, followed with frequencies and pick up locations. This day's flight lead/commander would be from the second flight platoon and there would be four slicks divided into four separate rotational lifts of two ships each, delivering a total of approximately forty-eight soldiers to their latest unfamiliar front line of combat. I had been assigned to flight lead of the second two ships. Bruce would be chalk two on the first flight and my chalk two would be from 2nd platoon. In addition, there would be a Charlie Charlie (command & control) ship overhead manned by 2nd Platoon directing us into the chosen LZ.

Normal mission planning routine, never the correct word in a combat zone, but we flew such missions on a regular basis. Maybe I should describe it as repetitive, but nothing was predictable.

The pickup zone was out in the flats, northwest of Titty Mountain where the troops had been gathered next to a permanent fire base. Arriving at my revetment we would be flying a different number ship on this mission, my regular bird tail number 041 was on this date scheduled for additional complex maintenance and accompanied by my crew chief Moody who would not be with us on this one, but gunner Dudley was still part of the crew, thankful to have him.

Preflight was completed and the time to crank had arrived. Copilot Thomason at the controls was maneuvering the ship from the narrow revetment as the rotor downwash battered the ship in its storage space, good training makes one sharp on the controls. We hovered out onto the peneprime strip (commercial composition of low-penetration grade asphalt and a solvent blend of kerosene and naphtha) between the revetments and were joined by our number two ship from 2nd platoon. I contacted the control tower to depart to the north. Thomason started moving the cyclic forward as the ship gave its aerodynamic shutter through transitional lift, mission on. Climbing north crossing the Ca Ty River below, another flawless morning, no hint as fate closed in.

We were on short final to PZ (Pickup Zone) as the lead ship of the second lift. The first two birds had already loaded up and the mission was rolling. So far routine, now again this is a word, when written, that almost always predicts the exact opposite is soon to ensue. How many stories in books, or testimonies, has included this harbinger phrase, a kind of teaser, but in real time during real events, it stays hidden in the mind, refusing to accept such thoughts.

We touched down to the hard-packed, sun baked, mud rice field of the pickup zone, now dormant until the rainy season arrives. The grunts (soldiers/paxs) lined up on both sides of the ships, ready to board. Those men were experienced, their uniforms were the first telltale sign, being faded and worn, their faces concentrated, showing their loss of youth as they now slowly loaded onto the helicopters, no dash here as they carefully located themselves on the empty cargo floor for their last chance to relax before their perilous arrival to the Bad Lands.

Flying troops into battle, you often look back into the cargo compartment, watching those men/boys unfettered enjoying the complete openness of the machine, their eyes searching the wilderness scenes below or tracking their fellow comrades in a machine flying close, usually less than two rotor discs in separation in a staggered formation, sometimes taking a personal moment to break out their cameras, snapping pictures of their friends sitting with legs hanging outside the machine as this magic ride speeds them to unpredictable destinies. The strength of a soldier isn't physical, its communal; placing one regular infantryman into a battle zone, I'm sure he would be terrified, but as a squad, his courage, competence, and rationale is multiplied by ten. That is the true potency of any Military.

Takeoff was smooth as the undisturbed air flowed through the blades, then we started our turn toward the mountains, our work load increasing as we mentally prepared our concentration to the mission.

First lift had landed, going in guns hot, one last suppression to give our men an edge. All was well as they contacted Charlie Charlie (command and Control) ship reporting they were returning to the PZ to

CHECK RIDE

onload more troops. Now our flight was in contact with the Charlie Charlie ship as we were moving northwest up a very steep valley, heavily encrusted with massive jungle trees. Charlie Charlie was directing us, guiding our flight up the valley as Thomason listened on the FM radio, moving our flight to those commands.

We were in the same area we had scouted the day before as the crew's eyes searched for the landing zone. Soldiers from the first lift had moved into the bush, covering down on the LZ to establish a secure area, they could not be seen as we approached. I jumped on the FM radio asking for conformation and exact location of LZ from Charlie Charlie ship as he informed us it was directly at our twelve o'clock position.

Thomason and I keep searching for any sign of the LZ, one was a very steep hillside covered with stumps, the other was at our twelve o'clock position, a small open lush green area just big enough for two ships. Most would think such areas are easy to pick out, but we have all read or listened to reports of large jet aircraft landing at the wrong airports which have rotating beacons with navigation aids and GPS coordinates, not to forget air traffic control towers. Finding a spot in the jungle mountains is not always a given. Most pilots pride themselves on having the eyes of an eagle, but even an eagle would have a difficult time with the jungles of Vietnam.

Once more I contacted Charlie Charlie to confirm our LZ location which he immediately replied, emphatically stating once again, "twelve O'clock", directly ahead. Thomason headed for the lush green LZ. We both agreed this must be the one, our chalk two had no comment and was just following as we were now on about a quarter of a mile final approach and would touch down shortly. I completed the pre-landing checks; my gunner and chief alert the troops thirty seconds as they pull their M60 machine guns to the cover position. We were going in cold, knowing our troops already occupied the LZ, the worry of accidental fratricide was always lurking.

Thomason was doing a great job; his approach angle was perfect, and he was shooting to the forward part of the LZ, ensuring chalk two had plenty of landing room. Dipping below the tree line as we slowed below twenty knots, the ship gave its aerodynamically shutter associated with loss of effective translational lift (Effective Translational Lift, or ETL) is a transitional state present after a helicopter has moved from hover to forward flight or the reverse. This state provides extra lift, most typically, when the airspeed reaches approximately 16-24 knots, but *is present with any horizontal flow of air across the rotor)*, seconds from touch down. The LZ was lush, but I quickly detected something ominous, parts of the landing area had been cultivated, my brain snapped into caution mode, we were about twenty-five feet above the LZ. Too late!

It had started; everything went into slow motion but was truthfully immediate as my peripheral vision seemed to catch all the events simultaneously. I heard the slapping of metal like someone hitting the side of a car door with their open hand as these shock waves rippled through the aircraft's skin. I saw the cyclic jump in Thomason's right hand as his arm tensed, adjusting for the unknown. I heard the clack, clack, clack of rapid automatic gun fire, maybe an AK. I caught sight of the soldier sitting on the cargo floor directly behind my pilots' seat jerk as a bullet ripped into his thigh from underneath, his arms automatically reached out defensively, grabbing the wound. We were taking fire and had been hit numerous times. The adrenalin surge was like John Travolta jamming a syringe into the heart of Uma Thurman from the movie Pulp Fiction. Life never seemed more important than when the thought of death squeezes your heart.

Thomason's reaction to the sudden cyclic input was reactive as his face quickly turned toward me. He had felt the feedback of what could only be bullets impacting our control system reverberating through his hand and arm. Quickly, as PC I grabbed the controls telling Thomason, "I have it!" I began pulling in power attempting to clear this death trap while scanning the instrument panel for clues to the ship's health. I could hear the whine of the turbine engine increasing, we still had engine power, but the large red Master caution light was illuminated with several yellow caution panel lights popping on, alerting us to some possible major damage. I was on the radio broadcasting, "We are taking fire!" as Dudley, my door gunner, was bellowing over the intercom, "Can I open fire, I have him." I quickly transmitted to Dudley, "Hold your fire, I don't know where the friendlies are."

Climbing out, the chief let me know chalk two was right behind us while we desperately were trying to make radio contact with anyone who could hear us transmitting in the blind that we have taken fire and had wounded onboard. There were no responses. Thomason was doing his job, our electrical inverters were out so he immediately went to the spare position, which also failed. We both were not panicked, too demanding for that, but knew there was damage, serious damage, while I still listened to the welcome drone of the turbine engine. "We are flying, we are flying!"

Maybe ten seconds had expired from receiving fire to our climb out as we now become aware of secondary concerns with the now distinct odor of JP4 jet fuel which swiftly filled the ship, our fuel tanks have been hit and ruptured, even with the self-sealing protection they were bleeding this highly flammable mixture into the ship's structure and the under belly of the helicopter as the explosive mixture of fluid was being pushed backward by our forward airspeed.

Multiple thoughts flooded our conscience...FIRE with a possible igniter from our damaged electrical system, not the way I want to go out; engine failure from possible damage to fuel pumps, crashing into the precipitous valley below; no smooth autorotation there, only the possibility of a mangled wreck falling through one-hundred foot trees. Also, how serious was our control system damaged? All the above with flashbacks already beginning, the soldiers screaming, cursing their helplessness surfacing as passengers, some just praying, the good ride had turned, this was what they will remember.

Flying down the valley, constantly on the radio informing my sister ships of our status, again with no response, maybe our electrical system had been so badly damaged we were not transmitting. Decisions/judgments occupying our entire thoughts, our damage was too severe to fly home. With the fuel tank damage, my first priority was knowing a fire or engine failure was a high possibility, and we would not make Phan Thiet, so where? I also knew at least one soldier had been wounded and needed medical attention.

Flying down the valley, continuously scanning my loss of engine options, but now just a few minutes away was the temporary fire base on top of the plateau, a secure area which afforded an option. My internal pleas of "engine please don't quit" persisted like some imp was sitting at my ear verbalizing this wish. Finally, someone acknowledged our radio messages on the net and we informed them of our intention to land at the fire base, some confidence returning, just needed the engine for one more minute. You had to tell yourself, "It's a machine, all the wishing, all the come-on baby, you can make it." Comments didn't change the fact that it's a machine stupid and it will fail when its reached its endurance, get it on the ground.

The plateau was now in sight, one last hurdle was a cliff on our approach, maybe three hundred plus feet high with large trees on its edges, but this was the place as I saw it, our only choice. Flying, the wind flowing through the machine, no discernable unusual vibrations, I chose a somewhat clear location to set the bird down, but we were coming in hot, no second chances; I had to plant it.

Working the controls, I shot straight for the ground, no hover, just that quick thump as the skids dug into the dirt with the nose rocking slightly forward. We started our emergency shut down as the troops unloaded, carrying their wounded comrade into the compound, never to be seen by us again, but sure he survived. Other troops came out from the fire base after watching our speedy approach and landing, thinking we crashed. Chalk two had followed us all the way and now he had also landed close by, with his status unknown. We were alive as the machine's blades slowed to a stop with that inner tension melting and we egressed our seats, the earth had never felt so appropriate.

High fives before they were called high fives as my crew nervously celebrated our individual fortune of war as we all started to assess the damage inflicted on the ship, eternity never appears close when you're young.

First and most obvious was the leaking JP4 dripping onto the ground, restricted by the self-sealing cells of the fuel system, no smokes now within fifty feet, that damage was caused by the enemy rounds that penetrated behind my seat, wounding that trooper with bullet fragments or ship fragments penetrating from the belly of the ship. Bullets striking aluminum/magnesium aircraft skin is like pushing a pencil through paper. We had armored seats and armored chicken plates, courtesy of lessons learned from previous wars, however as in most armored protection, there are always holes. Walking counterclockwise, we opened the electrical compartment panel which had been chewed up by rounds with the remainder of one round, like the magic Kennedy bullet lying on the deck. Thomason first observed this foreign object and picked it up. No doubt here, it was an AK-47 round where the steel core had been separated from copper sheathing. Thomason remained in possession of the steel inner core and I kept the copper outer casing, memento's/reminders which I still have these forty plus years later to memorialize our good fortune.

The UH-1 turbine engine sits on an upper separate deck, toward the center of the aircraft for center of gravity purposes. Opening both sides of the cowling, you could clearly see several enemy rounds had passed through the flooring, missing every vital component, POURQUOI?

The heart of the machine is its engine. There every integral component, like a human organ so sensitive to injury, congregates. Fuel lines, oil lines with reservoir's, hydraulic lines with reservoir's, numerous pumps, and large sections of the engine, compressor, combustion chambers, and exhaust section, any one of which, if penetrated by an enemy round, would most likely instantly cause catastrophic engine damage, ensuring our crash landing in that death zone.

Observing such damage, your mind now races through those possible scenarios as we continued our walk around, inspecting the right side of the aircraft. My God, right where my gunner Dudley was seated, a bullet had struck a seat bar which went vertical from the cargo floor, missing Dudley's head by a fraction of an inch, then continuing its unguided path through the aircraft skin into the transmission housing. The thought of losing anyone who you were responsible for was intense, especially only being twenty-two, but to lose a crewmember, my gunner, was penetrating.

Dudley and I nervously laughed at that close encounter of providence, once again too young to think life could end. Finally climbing on top of the aircraft, we inspected the rotor head components, instantly spotting a direct hit into a blade grip (blade grips, **which connect the blades to a hub, holding each rotor blade in place**). The bullet had penetrated directly straight in, about one inch, shaped about the size of a dime, on this substantial metal component constructed to hold the ships blades as they rotate three hundred and sixty-six revolutions per minute. The centrifugal forces placed upon this part are enormous and this damage could have resulted in metal fracturing, resulting in a slung blade. Every helicopter pilot since Igor is familiar with a loss of a blade story, and none of them turned out well. I visualize that moment my copilot Thomason's right hand jumped as this bullet impact transmitted through the rotor system controls directly to him like an electric shock, a sensation never practiced or reproduced in training.

Inspection complete, which had substantiated my initial presumption that we had significant damage. We were so fortunate that words stopped and life flooded in as I looked at the jungle surrounding our landing area, the sun now shining directly down with its warmth appreciated, we lived.

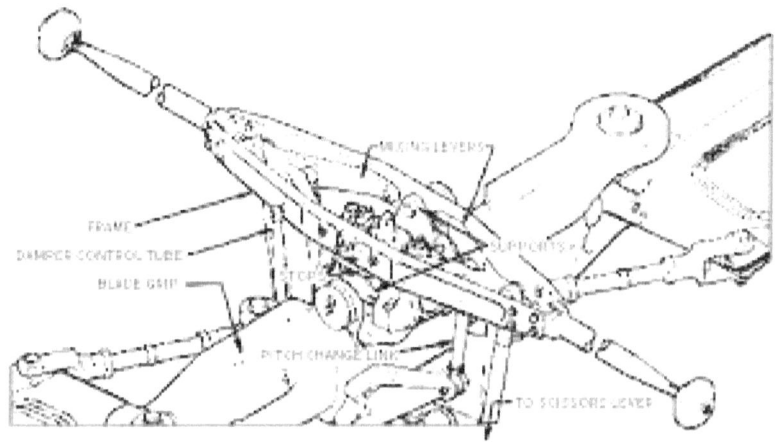

UH-1H blade grip

Vital information takes many paths, and our contact (military term for enemy encounter) had already been reported rearward to our Aviation Base operations back at the 192nd AHC with the basic details hopefully including our survival. From that instant, multiple problems had to be adjusted for, with the first being how to complete the mission with two of four birds out for battle damage. As I learned after shutting down, my number two bird in our flight also took at least one round up through the hell hole (inspection panel entrance large enough to fit a man) where many vulnerable components pass through. So they also were grounded until their battle damage could be inspected thoroughly by a technical inspector.

The insertion mission would be completed by Bruce and a sister ship while maintenance inspectors had been alerted and would be transported to our location for a detailed inspection of our ships for airworthiness. Our maintenance crew arrived maybe two hours after our event, deployed from Phan Thiet and consisted of a maintenance test pilot and an enlisted technical inspector. These are the guys who worked tirelessly every night as all the 192nd birds came home to be repaired and replaced, ensuring the Company had their eighty percent readiness level for the next morning's missions.

Upon their arrival, we immediately completed a second detailed inspection of the ship, still dribbling fuel on the ground as the bullet holes appeared bigger to me now. The maintenance guys rarely got to the field and, as I have explained, alcohol consumption was accepted back then and I could detect an odor on the team's breath. That fact didn't disturb me, they were very competent guys and even today having a beer during work is common in many occupations, maybe not aviation anymore, but those were different times.

Inspection complete, the maintenance pilot told me, "We will fly it back to Phan Thiet." Shock! Look I'm a PC, less than one year out of flight school with about one-thousand hours logged at this point, and there is absolutely no chance I'm flying this bird back as my instant non-aggressive response was, "NO-FUCKING WAY!" That was followed by my litany of the damage to the control system and the leaking fuel system. Now I still don't know if this was the maintenance pilot's humorous way of relieving my tensions, but with a hint of a smile he quickly said, "No, we'll hook it out." They started the arrangement process to have a CH-47 heavy lift Chinook Helicopter to sling load our damaged ship back to Phan Thiet.

CHECK RIDE

UH-1H Being Sling Loaded by CH-47

As midafternoon had arrived, my chalk two was returning to Phan Thiet, passing a flight worthy inspection, with only one round causing minimal damage. They offered us a lift, but I had already arranged for Bruce to pick us up after the completion of the insertion. A rigging team had arrived to prepare our ship for its ignominious flight being carried by big sister to home base. We had removed all personal items from the machine, also our two M60 machine guns with ammo. Bruce had finished and now landed and walked over while taking off his helmet, giving me his first words of wisdom with, "Hell, Buddy, I'm glad you made it." There would be less serious comments that evening.

Bruce was big brother and ready to take us home, a very welcome experience as I sat on the cargo floor with the doors open, like my soldiers only hours before, relaxing to the ship's distinctive signature sounds, knowing this day had passed. We arrived and I proceeded to base Operations, reporting this day's events with all happy to have us back, congratulating us on our survival.

Evening had arrived and I could see the CH47 deliver my ship to the pad in front of maintenance area. Days later, I would have an opportunity to speak to a senior civilian tech, an older gray haired gentleman, maybe in his mid-forties, whose explicit detail of my ships damage included that we had been hit by twenty one rounds of small arms fire and the ship was so severely damaged it would have to be sent back to the States for major reconstruction. As a professional tech, he knew how close to disaster we had come and smiled, thankful we were safe. I felt a confirmation of my first assessment and thanked him for his heartfelt concern.

The evening was commencing as the pilots pulled out their chairs, drinking beer or soda, smoking, joking but mostly concentrating on our good luck/fortune with their comments raging from, "Hey buddy, that was pretty dumb ass to get shot down," spoken by Thee Bruce, to "Can I have your food?" I'm sure that Norm had some worthy comical condescending remarks which enhanced the crowd's laughter at my expense while he postured with a cold can of beer and always the cigarette in his hand. Not one critique of this date's near disaster phased my fragile ego, I was alive and so was my crew, laughing was life's way of relishing this gift.

I had learned that we made our approach to the wrong LZ and that the LZ they used was the one with the steep hill and stumps, go figure. I wasn't upset upon this news because I had made every attempt to confirm the LZ with Charlie Charlie and this was not uncommon to happen in the jungle mountains. Often, even at flight school, close LZs had ships tracking to the wrong location, the significant detail was, we all survived. One really great ability of the helicopter is we can correct for such mistakes by taking off and relocating very quickly, unless you encounter the enemy.

It is almost time to retire and put an end to this date, with one last very important detail to complete, my after-action review in a letter to my father. Dad was the impetus for my desire to fly as a young boy, he would return from business trips throughout New York State as the President of the Police Conference of NY, with those wonderful colored maps and pictures, advertisements of the airliner he flew on. In the late fifties, there was a golden age of brightly enhanced advertisement brochures from the airlines, in today's age these are considered art. He had seen my interest in flying and took me on a cold November day to a small airfield in Armonk, New York to have my first ride with him in a tri-geared Piper Cub. Our pilot was sitting alone in front with both my father and me squeezed into the rear seat. I had some discussions with my Dad, and it was obvious he had a love of flight, but he was a depression youth and those times were tough, so dreams remained dreams. I was frightened with that cold nervous shaking during lift off until we climbed above the trees where the smell of the burning avgas mixed with the cold whistling wind over the wing surfaces, and my eyes scanning to look below, seeing everything like an eagle, imagined by a child of twelve or so.

Pure adventure, one of those first experience moments. Dad also paid for ten hours of flight instruction on my sixteenth birthday at Westchester County Airport in a Cessna 172 with a black flight instructor who was just so calming and informative, I wish somehow, he could know his part in my aviation path. Subsequently arriving to wear the silver wings of an Army Aviator these years later, I had completed one of life's dreams/goals and was most certainly guided by my father's hand and it would be to him I first confided my most serious event.

My letter started with, "Dad, IT HAPPENED SO FAST." Emotion tends to flow more in letters where your brain can view the written thoughts as I narrated to my father this day's happenings, which are now contained in this story. I did add, "DON'T TELL MOM!" Dad would die two years later, at sixty years of age, and I missed asking all those questions one thinks of after all the self-absorbed hustle of the twenties and thirties subside.

American Airlines Art Deco Brochure circa 1950s

Piper Cub

Sleeping that night was difficult as my mind flashed over with the day's events as the brain needed to correlate and store this memory. Of course, I was flying the next day, no doctors, no are you fit for duty, only more days of missions with three months to my rotation home and more new memories which would quiet down the previous ones. I wonder if my face now reflects that loss of youth.

Below is a report published and dated March 1970 that reflects a 192nd aircraft being recovered on the 25th of the previous month that was shot down and recovered with the crew chief wounded. Understanding that those who compile these facts do so months later and often are far removed from all the pertinent details. I believe this report is of my situation which occurred in late February on the 24th of 1970 where one soldier was wounded and mine was the only aircraft shot down from the 192nd and recovered by a CH-47 Chinook Helicopter during this period.

and one UH-1C destroyed, one UH-1H damaged, and one guard tower damaged plus four U.S. personnel WIA consisted of three hand grenades, two B-40 rockets, and twenty 82mm mortars.

While enemy activity was at a high point in the Dong Ba Thin area, the 155th AHC in Ban Me Thuot was relatively free of enemy attacks. Major support was continued for the 4th Infantry Division and other units in Pleiku Province. In flying 17th combat assaults during the month, Stagecoach aircraft also supported the 23rd ARVN Division and Detachments B-23 and B-50, 5th SFG operations.

The 192nd continued to support elements throughout the Phan Thiet region. The month started off with a combat assault for 1/50th Mechanized Infantry. The 192nd continued to support Task Force South and on the 17th, while conducting a combat assault for the 1/50th, had a UH-1H shot down by enemy ground fire. This action resulted in the door gunner being wounded. The aircraft was recovered by the 243rd ASHC. On the 25th, "C" Company, 75th Rangers made contact with the enemy and gave a call to the 192nd AHC for support. With guns always on stand-by for tactical emergencies, Tigershark, guns were launched immediately to provide support. The guns expended and were credited with three KBA's. The Polecats closed the month flying three combat assaults in one day. Aircraft were provided to the 1/50th Mechanized Infantry for two combat assaults and one for Binh Thuan sector.

The White Horse ROK Infantry Division commenced a two week operation on the 1st and the 281st was given the responsibility of supporting the 29th Regiment for the insertion and resupply. In support of the 23rd ARVN Division, the Intruders flew eight regimental combat assaults throughout the month as well as combat assaults for Mike Strike Company, 5th SFG.

42

28. Short Timer

Short timer

/noun/

1. A person, as a soldier, who has a short period of time left to serve on a tour of combat duty.

End of April 1970 and cruising into May, one month plus to end of tour, feeling time wearing me down. Reaching this stage of now being one of the senior guys, all have forgotten your name, and your call sign, now referring to you as just "hey! short timer". Good and bad reflections about this moniker. Since my arrival eleven months before, many normal conversations always included a soldier's wish, one which may span centuries. On multiple occasions, the topic of one's possible demise was discussed with all in agreement that one would desire, if fate was to dictate, to get "IT" quickly. IT being death in battle, selfishly said to avoid the next twelve months of perceived impending bull shit, no honor, no glory, just that raw genuineness of reality. Now such statements may be comical in nature, but still these errant types of thoughts exist.

Somehow, I'm sure everyone has experienced this variety of self-pity which miraculously helps one to think through the problems in your immediate future. Most young men aren't really forward-looking individuals at this age, in my opinion. One never thinks you will be forty in this business with the flying of daily missions into enemy territory. These contemplations are reinforced with the daily news of some aircraft being shot down or crashing in country. Now having arrived at this epoch and avoiding this soldier's wish, unfulfilled by the grace of the God's, one has entered the reverse scenario where you now have the desire to exist, knowing and thinking that your death at this late stage would devastate your family, perplexing them with the why make him go through all of this drivel to fall short.

Truth is random makes you question grand design, but keeping it shallow, the use of my new name does make one start to think a little too much. It's comforting to observe your own shadow, knowing it has showed up for this day's flight, a good omen. I had one month of combat flying left and loved being a functional cog, always enjoying this camaraderie but ever aware of the downside, like a dog given a task, happy to be an asset, a part of the family but knowing our life span could be shorter than most.

News had arrived that President Nixon was ordering troops into Cambodia, on April 30, 1970. WHAT!! Or in the new age language of 2012, WTF! We all know what this Presidential revelation means a very strong possibility our unit will be tasked to participate since the NVA's main headquarters on the Cambodian border is due west about 248 miles, a full tank of gas to a very inhospitable area.

Looking back through these many years of our country's military excursions, I often wonder why on so many Presidential public broadcasts we, our nation, chooses to enlighten our enemy on our real-time intentions rather than wait to post announce these decisions to effectively use the benefits of surprise. Does the public not fighting at home really need to be sitting right on the bench? The enemy has an extremely robust intelligence system, maybe worthier than our own. How many occasions have we seen Vietnamese's employees pacing themselves from a tall LZ feature like our water tower, X + number of paces in such a direction could be a fairly accurate measurement for future placement of enemy mortar fire.

We had noticed increase activity near our base while now conducting a new mission, an evening aerial observation flight around our perimeter of LZ Betty, called fire fly. On one occasion, I had hitched a ride, sitting with my feet dangling from the cargo area holding an M16 with a full clip as the ship slowly maneuvered forward at sixty knots at about one-hundred feet in altitude, flying over the area below where most all the attacks on Phan Thiet/LZ Betty originated.

This was not unusual for pilots to include themselves on such flights since variety helps keep ones' attention and this was fun sitting on a movable perch scanning the below terrain as the evening air rushed completely over you. Who gets to do these things as the hues of the oncoming dusk begin their movement like a tide over the day and your little boy comes out? As we flew about a mile from the south perimeter, we observed a large group of men in the low savannah, unarmed, moving eastward, all dressed like villagers. I mean it was obvious they moved in unison and were directed, but the PC was told to just monitor and report their movement to operations without engaging them.

This close to base you didn't have to be the CIA to understand something was unusual. Our reports were forwarded to others to compile this intelligence. We landed as the grey sky turned to the dark grey before night fully encompassed the base. Back at our little courtyard and during our nightly deep thought discussions boosted with a variety of beverages, a collective assumption amongst the pilots was we were going to get hit, and soon.

Throughout my eleven months at LZ Betty (AKA Phan Thiet) we would receive mortar and rocket fire on a regular basis. There was one earlier occasion that this type of attack was extended and a decision was made to have all the air crews' standby their respective aircraft for possible evacuation. This evacuation plan was formulated to deal with the potential of the enemy completely overwhelming LZ Betty and destroying all the aircraft in their revetments. This detailed course of action, which all pilots and crews were familiar with, was that last minute last ditch event, (This **expression** alludes to the military sense of last ditch, "the last line of defense) but I often would think of what don't we know that our upper command structure has put into place a preempted plan specifically devised to mitigate aircraft losses. Talk about being left out, how serious is it that they want to evacuate all the ships into the black night to avoid being destroyed in their revetments. Crouching in the dark next to the helicopter and heavily armed, your mind in its spare time dwelling in the darkness eventually realizes that if we deploy, what happens to all those who stay behind, unable to get out, and what must they be thinking as they watch the only **life boats** leaving **empty**, a true Titanic sensation, so I will call this operation Captain Smith **(Edward John Smith, Master/ Captain of the White Star Vessel *Titanic* who perished with his ship).** Once again to rehash…it's good to be in aviation.

CHECK RIDE

3 MAY 70 0145

```
LZ Betty began taking incoming. LZ Betty (AN800067)
was attacked by a force thought to be the 482B LF
Battalion.  The attack consisted of approximately
125 rounds of 60mm and 82mm mortar fire, an unknown
number of B40 and B41 rounds, and sapper probes in
four separate locations of the perimeter.  All five
companies of the battalion appear to have partici-
pated in the attack. 5 LZ Betty received 100-125 rounds 60mm
and 82mm mortar, unknown B40 and B41, unknown satchel charges and
simultaneously a sapper attack. 6 KIA, 35 WIA, Enemy 14 KIA. 4
```

A **battalion** is a military unit. The use of the term "battalion" varies by nationality and branch of service. Typically, a battalion consists of 300 to 800 soldiers and is divided into a number of companies.

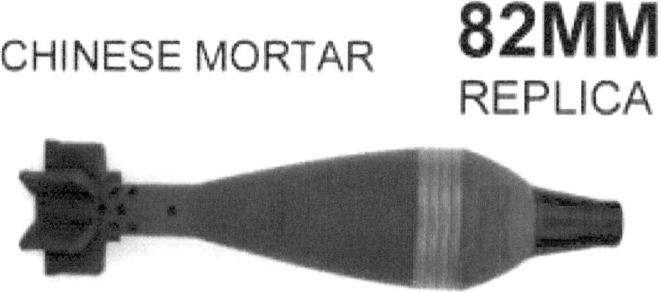

82 MM mortart round

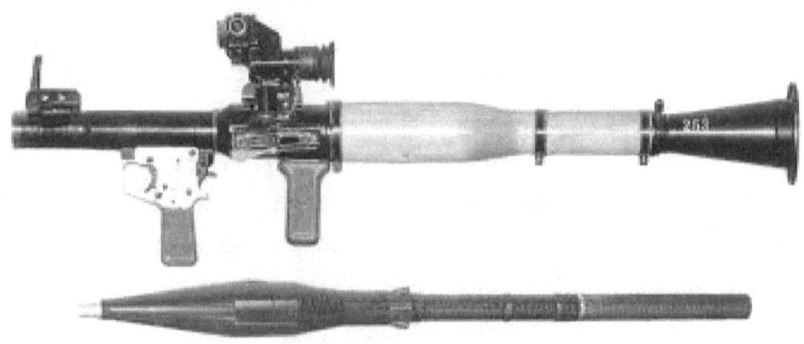

B40/B41 Rocket Propelled

May 2, 1970 end of day, all settle in for the usual drinks and bull shit sessions. Sitting outside our hooch's, some pilots were swinging in their nylon hammocks. Our Army family/brothers gathered to celebrate one less day on our Julian Calendars. Normal routine, dinner, showers, mission board checking for tomorrow, then off to sleep. It still amazes me to know how quickly one can wake up from that low pitched wrump, wrump, of mortar rounds impacting and this resonance creeping closer.

Early hours May 3, 1970 as the warning siren began to scream, it was obvious numerous mortar and rocket rounds were being directed at our home. This distinctly shocked all out of that deep sleep, the kind that repairs the mind. Wide awake now as this new day has arrived and the incoming mortar and rocket round pace aids to accelerate our chaotic scrambling to dress while grabbing guns and equipment as the platoon leader begins to bellow out the order to go and standby the ships. Never good news this early in an attack as the base has erupted into serious defense mode.

CHECK RIDE

Running in the dark, trying to maintain your footing over familiar ground in daylight, but now every leap is a possible broken leg with your feet feeling the way forward. Every defense bunker and tower now pouring rounds downrange. Regardless of the enemy's focal point, the noise was engulfing as your hearing seemed to capture it all. First indication of a real threat was the appearance of green tracers flying sporadically overhead, knowing each gap contains four rounds in between, unseen but expected, followed by the eyes peripheral ability to view our own defensive outbound red tracer trails, from every post on our perimeter, colorful, with red and green steel bullets now mixed with the sporadic white-hot fragments of mortar rounds, all visible on our sprint to the machines. Sometimes young and stupid works to your advantage.

The US Army uses red tracers, so it was immediately obvious the enemy was within small arms range as their green tracers and explosions continued. Every perimeter tower and bunker was firing with their tracer's interweaving, protected by their enfliade position. Numerous explosions with their thumps followed by concussive pressure wave even felt at some distance. It was like sitting under some fireworks display, but with steel splinters as fragments.

When used, tracers are usually loaded as every fifth round in machine gun belts, referred to as four-to-one tracer. Platoon and squad leaders will sometimes load their magazines entirely with tracers to mark targets for their soldiers to fire on. Tracers are also sometimes placed two or three rounds from the bottom of magazines to alert the shooter that their weapon is almost empty.

Tracer rounds may also ignite flammable substances on contact from a nominal distance.

Reaching the aircraft revetment, you felt that flood of reassurance and the familiar perception of safety in our defensive position, a fixed place to make a stand as we found our breath from the dash. The revetment was maybe fifteen feet in width, large enough to maneuver a helicopter into. The protective walls were composed of steel double planking filled in with earth, creating this slot and about five to six feet in height allowing each crew member to cover down on each corner.

Helicopter Revetment

CHECK RIDE

THE ATTACK ON LZ BETTY, MAY 3, 1970!

Edited by Ray Sarlin

Webmaster's Introduction *On April 30th, 1970, President Nixon announced a U.S. and South Vietnamese "Incursion" into Cambodia, triggering a wave of protests across university campuses in the United States that culminated on Monday, 4 May, with the death of four students at Kent State University. While those well-known activities were screaming from the media headlines around the world, a desperate battle was being waged at LZ Betty , Phan Thiet, Binh Thuan Province that will never be forgotten by those who were there, but will never be known by those who weren't. At 0145 hours on 3 May 1970, LZ Betty, the home base of IFFV's Task Force South and the 1st Battalion (Mechanized), 50th Infantry came under intense mortar and rocket fire, and five companies of VC sappers attacked the wire, achieving penetration in two places. The men who fought off this assault were mainly the battalion's rear troops from Headquarters and Delta companies, who were no less brave (or confused) than line infantrymen were in battle, and whose blood was just as red.*

Copyright 2003, www.ichiban1.org, Ray Sarlin and Contributors as noted below. All rights reserved.

Accounts of this confrontation by some have been called a battle and I would agree. Both a cake and a cupcake are made with the same ingredients, only the size will be different. Men will die this night on both sides, not in the numbers of Iwo Jima of WWII, or in the battle of the Chosin Reservoir during the Korean War, but in a scaled down version of indiscriminate war. The emotional pain to the victims' families of this night's raid will be none the less painful as we sit and watch this conflict erupt. Intensity of the battle now picking up, still numerous mortar and rocket round impacts as we look south toward the location of the 2nd of the 7th Cav, 1st Air Cav Div. The Cav has maybe six Cobra Helicopter gunships with nine UH1H troop transport machines and about six OH6 Cayuse helicopters, a smaller machine used very effectively as an airborne scout. The Cav Unit is just over the rise on the very south tip of LZ Betty but we hear the explosions and clearly see the glow of fires flourished with some powerful accelerant. To the southwest we see one of the watch towers on fire; the enemy has penetrated the perimeter. This fact is almost unbelievable there are multi layers of barbwire which crisscrosses in a pattern a spider would be proud to call their own. The perimeter was a cleared area and my imperfect description which I will give you is that it looked like part of the no man's land from pictures of World War 1, empty of trees, plain, pock-marked from shelling over these few years of LZ Betty's existence. Followed with irregular terrain and further out the headstones of the graveyard, no rest for those souls. The Vietnamese grave yard on the west side of LZ Betty had now come alive with the enemy using these death monuments as the perfect cover to launch this attack. Tangs from all the weapons and explosions now drifted across the runway into our revetment area, the perfume of war mixed with patches of blue smoke. There was no discernible regularity as the battle continued with scattered explosions and the pulsing clatter of multiple caliber weapons. Our eyes strained in the dark, searching for possible sappers. Our vision was assisted by the occasional illumination of artillery flares which were fired overhead on the edge of the west perimeter, just across from the airstrip.

Those flares floating down on their small parachutes radiating that very eerily amber glow which frolics one's mind with the wavering shadows across the revetment area, now being aware the enemy had breached our perimeter. That made all those shadows ominous. Kneeling on the edge of the revetment, time slowed as you watched and listened to the ongoing battle while holding your weapon, ready for what may come. I had an M2 carbine with six clips of ball ammo and my trusted S&W 38 pistola, all had been charged with a live round.

Over the next hour plus, the sporadic explosions and gun fire naturally reduced but supporting incoming 105mm howitzer fire from the American fire bases Sandy and Shari located to our west were pounding the grave yard. Imagine our own artillery guns aimed in our direction, whoa! A lone Cobra Gunship from the Cav Unit had managed to launch and now made its gun runs from the sea side of LZ Betty to the grave yard, firing its minigun directly overhead of our revetment, expelling hundreds of brass shell casings down onto our area. The sound was awesome, like being seated directly in front of a huge woofer speaker vibrating every cell in your body. The roar of the minigun combined with the power of the Cobra's rotor blades pounding the air made you proud to be an Army Aviator. The red stream of death from their guns raked the graveyard as the rounds ricocheted back into the night in a chaotic splash.

About two hours had elapsed as Spooky arrived, an Air Force C47 converted to a gunship with four miniguns combing the outer fringes of our perimeter as the fight had receded, chasing the enemy retreat. Finally, the word was passed to stand down and return to the hooch's. All was safe. The truth is, this attack had killed six Americans, wounded thirty-five, and sappers had destroyed three helicopters from the Cav in their revetments, safe is subjective.

Cobra Gun Ship

One month was left until I leave and over the next three days we all learned of the bewildering encounter at Kent State University where four students had been tragically killed over the protesting of this war. We all felt regretful about those students' fate on the same day that we were under attack. My sister Janet, senior child of Tom and Theresa and future matriarch of the Yonkers McGurns' had related when I arrived home that there had been a small article in the newspaper stating that LZ Betty was attacked and she had inquired of my parents "isn't that where Tom is", but then again the headlines were about the protest, not the six lives lost and thirty-five wounded that night. History always verges to the chroniclers' recordings and their perception of prominence.

> **S&S Vietnam Bureau**
>
> SAIGON — North Vietnamese and Viet Cong gunners unleashed 77 Vietnam-wide rocket and mortar attacks late Saturday and early Sunday in the heaviest round of shellings in more than a month, official spokesmen reported.
>
> The attacks were the heaviest reported since the 24 hours covering the night of March 31 and the early morning of April 1 when Communist indirect fire attacks hit Allied bases and towns 130 times. At that time, ground attacks were coordinated with many of the shellings, but this was not the case Saturday.
>
> With one exception, ground action throughout Vietnam Saturday was light. The exception was a sapper attack on Landing Zone Betty outside Phan Thiet 110 miles east of Saigon early Saturday in which five U.S. troops were killed and 36 wounded. The Task Force South base, according to a spokesman, was partially overrun, but the enemy was repulsed after losing an unknown number of dead.

US Newspaper Article Mentioning LZ Betty Attack Phan Thiet

Walking back to the hooch, the night soon would be replaced with that light grey sky announcing the new morn. As the new sun's brilliance predictably moved westbound, we would all be flying presently on the new day's missions while the smoke of the night's attack still filtered throughout LZ Betty. One long day shorter.

29. Finishing it out

The last month has arrived, even though it has been ushered in by a substantial attack on our home base where six soldiers were killed and thirty-five wounded. Now in the second week of May 1970 as marked on my Julian calendar, I stared at the filled in boxes, like Edmond Dante's imprisonment at Chateau"d'if, hoping for my escape. Just rhetoric, as deep inside I wanted to stay, I wanted to fly on missions thinking always to support our troops, and I enjoyed the team. Other problems had surfaced, never anticipated, this pace was now showing its toll on my physical being. Upon arriving in Vietnam, I weighed 220 pounds at 6'3" and now I was the thinnest I had ever been as an adult, weighing 170 pounds. The constant flying on most every day had furtively worn me down.

There was one instance in these last two months which was disturbing. I was hitting the sack fully expecting to fly in the morning but sometime during the darkness I was struck ill. It seemed I woke in a warm sweat with pain radiating throughout my entire body while simultaneously being paralyzed and even unable to call for help, incapable to move any portion of my substantial frame. My conscience leaning to a diagnosis of possible food poisoning as pain racked my body coupled with delusions of my condition now completely taking over reason. Bruce, my roommate, had been gone several days, reassigned to fly with the 92nd out of Dong Ba Tin, so no chance of assistance there, as for the first time in my entire life I experienced that I may die this night alone, as the pain overwhelmed my self-pity while frozen on my bunk. The darkness seemed protracted as the hours passed when, finally, I could slide off my bunk, forcing myself to move toward the door hunched with abdominal pain and sweat. No warning as my bowels felt like I was being eviscerated, I thought my entire intestines were exiting my body as I unceremoniously soiled myself. I could not even comprehend that I had no control but was way past the emotion of embarrassment. We all got sick and caring for each other was the norm. Dragging myself to the showers, now empty, I stripped my clothes, disposing them into the nearest trash barrel, then quickly entered the shower filled with those dark shadows from the wooden framing, fully expecting the unheated water to cleanse. As the shower spray hit my body, it had that immediate cooling effect of reducing what I perceived as a fever, while restoring some mobility to my cramped corpus. Returning to my room, I hastily got new shorts while trying to clean the remnants of my biological mess, but the episode was not complete as I kneeled in front of my hooch door as the body was not convinced that the demon was out. I continued dry heaving until the grey light arrived, utterly drained of all energy, wishing Mom was there touching my forehead as when I was a child, reassuring me all would be fine. 1st Platoon pilots were now waking and showing their concern, (always welcome) as they paraded on their way to today's missions. Capt. Boley came over after scratching my name from the flight assignment board, ordering me with genuine compassion to go see the Doc at 7AM. I was first in line and after explaining my long humiliating night, the Doc gave me the smallest pill I had ever seen and ordered me to bed rest. Returning to the hooch after ingesting this miracle pill, I was placed into the deepest sleep I had ever experienced, not waking until almost twenty-four hours later, I wish I had the Doc's name. I never really recovered from this event until after my arrival home.

CHECK RIDE

At this stage with approximately three weeks left in country, you were looking forward to receiving the military's version of a ticket home, "Orders". Several weeks before, I had been instructed to complete my dream sheet, a piece of paper informing the United States Army of my desire for future assignments. My choices written in order of importance were Germany, Alaska, and Korea. I was hoping to continue my flying career and see the world, a young man's wish. Two weeks left as my WONKA ticket arrived. I had been ordered to Fort Rucker, Alabama. Christ! Fort Rucker, really! Haven't I had enough of that place with all the anxiety of the past returning? Time to soldier up, you are not in charge of the events in your life, you are only in charge of your life, "make it work".

All of us anticipated an early release from Vietnam, some pilots received orders taking three weeks off their time in country, and mine were dated to depart from Cam Ranh Bay on June 1, 1970. "THE FULL MONTY", three hundred and sixty-five days! Capt. Boley now started to assign me the bus runs with the new copilots. It was his way of making sure I had the best probability of survival now having arrived this close to my end of tour, a practice all appreciated.

After my illness, I had never fully recovered and during a two-ship mission to Dalat, a most beautiful French influenced City in the mountains, I had what most would consider a very unusual occurrence. We had landed at the lake pad where locals would bring their work elephants for a bath. Our version of a drive through car wash. The mission was to deliver then wait for some hours while our military passengers completed their assignment. As in the real world, this mission could have been requested under the veil of military importance when the requesting unit just wanted to give some young officers a day trip, "it's all in the writing of the mission request".

Between our two crews, there were a total of eight and Dalat had a respectable reputation for attractive young women with the mixture of Vietnamese and French qualities. As the cumulative majority of our crew discussed a place, they knew which could satisfy their **"need"** for **"seed"**. I, along with two other crew members reluctant to participate (I'm still trying to recover), now chose to just stay together as a crew walking through the winding hill streets of Dalat, beautifully lined with wildflowers and that humid vegetation scent. It was rumored that the enemy also often would infiltrate into Dalat for the same reasons, their version of R&R (Rest & Recuperation) in an open city. Wandering through these streets as eight armed men was comforting. We arrived at a beautiful stucco residence, definitely French influenced, and landscaped with that exotic jungle type vegetation. How enlisted guys always had intimate knowledge of such places is an amazing conundrum. Never underestimate the enlisted guy, a true source of other possibilities.

I was now with my fellow non-participants inside the residence, seated in a completely open room with several unselected young ladies waiting for the remainder of our crew's completion. It hit me like a bolt of lightning, some giant foot stepping on my colon as I instantly plead to these young girls for a bathroom. Believe it, most of these young women had the basics of the English language mastered, and obviously recognized my situation by directing me down one floor to the toilet. It was not the low point in my life, but memorable, as I arrived one floor down and there in the middle of another completely empty room was a porcelain flush toilet, no walls, just a white monument. What was the builder of this home thinking, almost a miracle in third world countries?

The plumber who installed this had an easy job. The Catch-22 was the room also had several young ladies seated against the walls; this must be one busy business. It is, what it is! As I proceeded to disengage my trousers, mandated by whatever bug had penetrated my system, and hurriedly take my rightful seat on this throne with a 360-degree view in the center of this imagined harem. High pitched laughter from young foreign girls erupted as these ladies endured my embarrassment. All I could do was join in with a stressed stupid smile accepting this odyssey.

I wonder how the great Greek poet Homer would have spun this tale and what moments he used as inspiration. I thank all the basic training latrines where there were no walls, as soldiers sat four across, knee to knee, making deliberate conversation in preparation for that moment, Army Aviation at its best.

US Army Basic Training Latrine/Toilets 1968

Capt. Boley stopped scheduling me one week before my departure, more insurance that I would make it home, time to adjust the mind for reunions. During the last two weeks of flying, I was cognizant that our flights in this country were totally unrestricted and I wanted to enjoy this sensation completely before my final freedom bird ride home. I took advantage of such flying opportunities often while coming south through the Phan Rang pass then continuing toward the beach, now turning right and descending to a low-level flight profile (flying below fifty feet) above the beach. This is a sight rarely seen in the US, empty pristine beaches for fifty mile stretches as the South China Sea waves continuously churn the sloping yellow sands, no signs of humans, with the smell of the moist saltwater cooling our ride. We (the crew) were just another gull enjoying the freedom of flight above the waves.

There were other incidents during this last hurrah of flying, one during a bus run where I spotted an overloaded bus on Highway One going north-bound kicking up a dust trail on the much-worn highway below. It was something which you might see pictured in a magazine, with people and their possessions on the overloaded roof of a multi-colored bus. I couldn't help myself, I'm going to buzz it and give them a thrill, but now realizing the thrill was for me. Not an out of envelope maneuver, like the Duck of the past, just a haircut, as I proceeded to lower the collective while informing the crew what was to come.

If you have never had the opportunity to be directly in front of an oncoming UH-1H helicopter low level, you first will hear the noise of the pounding blades followed by the vibration building as it swoops directly at you, "awesome sound". I was aligned with the road and closing fast on the bus as I clearly saw a young boy, maybe twelve, seated on top of the bus as he quickly removed his shoe and tossed it as we passed ten feet over the top of his ride. His shoe harmlessly struck the bottom of our machine. I wanted to land and applaud him; he was beyond his years in courage. The Vietnamese are good people, no matter how this war ends.

Another rare mission occurred when I was assigned to the bus run flying up to Dong Ba Tin. Here I would pick up a Philippine Rock Band who were scheduled to entertain the lost souls of LZ Betty. I can only remember two shows at LZ Betty and the other one was an Australian Rock Band. Such small entertainment bands often travelled through the small bases, sponsored by the USO. We heard Bob Hope once flew over our base on his way up North. The true entertainment was both these

Bands had Go Go dancers. Young women dancers in bikini type costumes gyrating to rock music on a flimsy wooded stage while soldiers drank warm beer screaming like it was Woodstock.

Now assigned to transport this group, they were all onboard and the crew chief and gunner finally had passengers that didn't smell like jungle. Their aroma was that of cookies and cream. The girls were excited and flirty, but hey, that was their job and they were good at it. While airborne, the chief escorted one of the girls between the seats as she started to step over onto my lap as I was flying. The copilot took the controls as this woman, hell she could have been older than me, sat directly on my lap for a few minutes before returning to her seat. She was unaware that her actions would most likely be our courtyard center of conversation that night after landing at LZ Betty. I have slides somewhere of this group on stage. While writing about this event, I hope the US Army statute of limitations has expired.

One of my very last missions was a bus run flight to Dong Ba Tin, our battalion headquarters. It was the place I started this odyssey eleven months before and now would be transporting a mirror image of myself, maybe even my replacement, to their new home at Phan Thiet. Our flight north was uneventful with several stops at outlying posts for other possible soldiers on their way home. Now topping off with a full bag of gas we began our hundred and fifty-mile trip back to Phan Thiet. My copilot was flying paralleling Highway One, maybe at five hundred feet, "or lower", as we enjoyed the lush emerald green landscape below...numerous rice paddies with men working their prized water buffaloes in the chocolate brown waters surrounded by lush grass covered mud walls. Flocks of white birds disturbed with our approach were taking to flight, peeling off left or right, contrasted by the palette of ground cover.

Unexpected and with no time to react, my copilot was flying straight and level as a large, no very large, crane type fowl was directly in my focus. In an instant, this flying body of flesh, bone, organs, and feathers crashed into our left side wind screen, it must have been four feet across. The impact flexed the very thick aircraft plexi-glass windshield inward as I sat staring with my visor down, about to eat raw bird, once again hoping this doesn't hurt. It was a miracle the windshield held as the relative wind of our forward speed thrust the carcass over the top to be diced by our version of an oversized Cuisinart, the rotor blades. Hazards of nature, just another experience to round out the year.

My last few days I was grounded as "thee" shortest short timer in the company. No more letters home since I would most likely at this stage arrive before the mail reached my home. I had written all to stop any correspondence, reassuring those I loved I will see all of you soon. I watched as my fellow pilots left every morning on missions for our troops. I reminisced of those holiday runs of hot meals for our troops in the boonies, flying all day for them to celebrate Thanksgiving or Christmas, enjoying my part in their small remembrance of home reinforced with hot turkey in thermite cans.

I wandered the company area looking closely at my faux home, realizing I most likely would never return to this place halfway around the globe. Two days were left on the Memorial Day weekend as I packed up, leaving my share of essentials behind for the "Bruce", including our prized possession, a small refrigerator. You can't appreciate such an appliance until you don't have one, a cool drink is worth a gold bar.

The last night in Phan Thiet/LZ Betty as all those I had flown with toasted my departure and presented me with my Plaque inscribed with my call sign "MOTHER".

"I made it" as memories of those we lost flashed in my mind. Especially John Wright, I can still picture him at White Hall street on our beginning. One positive Capt. Ken Boley had assigned the perfect crew to fly me north to Cam Ranh Bay. At the top of the flight board was PC Bruce Britton and PC Norm Niswonger acting as his copilot now assigned as my personal pilots. What could be better than my two best combat brothers flying me on my first leg toward home, a feeling of immense serenity, they would have been my choices. I was ready to sleep my last night in LZ Betty, completing the obligation I had volunteered for. Peaceful as I dreamed of coming reunions.

I awoke late, maybe 7:30, dressed and walked to the mess hall for my last meal in Phan Thiet of eggs over easy with over cooked bacon, toast, and home fries. Bruce and Norm were on an early mission and would pick me up around 11:00. My bags were in front of the hooch as I talked to Ken with that 'see you in the States' type conversation.

My time had arrived. I was at the ship, my bags were tied down, and I was sitting in the right cargo seat on the edge with the helicopter doors pinned back as we takeoff. I recognized every feature on the trip north as my eyes for the last time viewed the sea, mountains, and Le Hong Phong Forest below. Physically the clues had finally revealed themselves. I was exhausted, not just fatigue but chronic fatigued. I could let go, my mission was over. Before leaving LZ Betty, I had visited the Doc, afraid I wouldn't be able to sleep on the way home and he issued me a half dozen pills, just in case I needed something for the long journey.

Landing at Cam Ranh Bay and shutting down as my last man hugs with Bruce and Norm took place, of course more than the usual jabs as my last umbilical to the 192nd prepared to takeoff, and I was concerned about them. I reported into the transport center whose minions escorted me to an open barracks filled with empty bunks, no sheets and I could only wonder my mirrored arriving remarks of what had slept here. Awaiting further instructions, I had been informed that this stay could be a few days. I sat on a bunk sorting through my baggage, thinning out the load of what I really wanted to carry home. I left most of my jungle fatigues and brand-new jump boots. I wanted to travel as practicable as possible, and urgency was beginning to set in.

Surprised, I had been there about ten hours and got my flight assignment to leave within the hour, it was happening, the reverse journey.

30. Reunion

Rolling down the long runway onboard my freedom flight as the DC8 aircraft rotated its nose, angled steeply, gaining sky and a cheer/roar from every combat brother on this flight echoed through the aircraft, "We're out!" as Cam Ranh Bay sets in the west. I had been worried about sleeping soundly in my last weeks, but I quickly slipped into a deep sleep. The aircraft and distance from Vietnam were now my blanket as we moved eastward. My mind had taken over the repairs.

The journey was an extended one and I do not recall a fuel stop, but perhaps we stopped at Guam then continued on to Ft Lewis, Washington, my original departure point twelve months before.

Bizarre were my thoughts of backtracking this treaded route. Landing we disembarked with all on board now having to process through US Customs and as I moved through the line, I declared my handful of pills given by the DOC. The customs agent acknowledged my sincere honesty, or maybe it was my subdued appearance seeing my oversized clothing, as he waved me through. How many young men he must see with the faces of the lost?

One young soldier from our flight who was proceeding through this customs point was unexpectedly shocked when the agents discovered numerous pills, maybe one hundred plus in his watch case. All soldiers did not participate in combat. Many were assigned to administrative positions in large cities like Saigon, a place where addictions could easily be acquired. The soldier was now politely requested to exit the line as a new journey begins for him, one not expected.

Leaving customs, a uniformed E-8 Master Sergeant met all with a firm double hand handshake welcoming us home while directing us to the mess hall where he assured all that their steak dinner awaited. The steak dinner was a tradition and although appreciated, this token meal had no meaning for me as I asked the sergeant if I could just leave the base to start my trip home. The Master sergeant smiling said, "Sir, you're an Officer, you can leave anytime you want." I was out the gate, catching a cab to SeaTac Airport, mission get the fuck home.

It was late afternoon as I checked into the airline counter requesting the earliest flight to JFK and discovered it would be at 11PM. I purchased a ticket for the red eye (A **red-eye flight** is any **flight** departing late at night and arriving early the next morning) flight then proceeded to the nearest motel across the street from the terminals. That must not have been an out of the normal occurrence as I explained to the manager I only needed the room for a few hours, whereas he presented me my key.

AHH! Heaven…my own porcelain toilet with walls and a tub, air-condition room with a black and white TV. Junk food machines in the lobby as I stocked up, preparing for my first real bath with unlimited hot water. After a quick nap, the alarm clock made sure I was up at 9PM Pacific. Changing into my civilian clothes, of course nothing fits, I was so skinny, and I didn't give a shit. Catching my flight, the plane was practically empty as a very kind stewardess directed me to a row toward the back that was completely unoccupied. She must have seen a lot of us, God Bless her.

Arriving JFK at 6 AM, Eastern standard time, the sun was starting to peek over the horizon, a time of day I have become very accustomed to. Exiting the airport, I flagged a cab, now in a brief exchange with the driver to take me to Yonkers. The driver was a character from every movie I had ever seen with stereotyped NYC cabs drivers. He began his interrogation, inquiring where I had come from and the universal phase of "how was it" after my modest reveal of Vietnam. My driver was also an astute listener, possibly a condition of caffeine induced energy as he peered many times in his rear-view mirror as if watching me speak enhanced his comprehension. Naturally I had my safety concerns, desiring for him to maintain the vehicle on the highway. Imagine almost home as the cab runs off the road. The ride cost me twenty dollars with the tip, always good to remember the cost of services in the past.

It was around 8 AM as I arrived at 2 Berkley Ave where Janet, my wife, resided with her family as I knocked on the apartment door, unannounced. The reunion was warm as her mother Meriam insisted I have breakfast, all stunned at my arrival. I could see they had noticed my slender appearance but were polite in their approach.

Using their old dial phone, I called my Father. Dad was at work as a Police Sergeant with the City of Yonkers as he dropped everything and rushed to see me. I exited the building, waiting at curbside as I watched my Dad's car approach. Speeding toward me, it was clear he had missed me. Dad jumped out like a stuntman from the barely stopped vehicle and I hoped he had placed the car in park.

As Dad moved sprightly in my direction, his expression was "OMG Tommy looks terrible," but his hug was the embrace of a father as years of being his son returned a spirit to my body. We don't ever get enough of such moments.

Dad had called Mom, who was at work, employed as a secretary in Refined Syrups Corporation. As only men can do, Dad explained for Mom to meet him at the foot bridge crossing the railroad. Dad didn't tell Mom why, but somehow word got to her co-workers on what was about to take place. It was my mother who managed to get me a summer part-time job right after graduation at Refined Syrups. In 1967, I was paid $6.10 an hour, more than decent wages for a teenage boy at that time.

I could see my mother walking up the steel steps from the work yard below while her co-workers were staring out the windows behind her. As she started to cross the bridge over the train tracks, she was now seeing her son. Her hand went to her face as she hurried her steps as I rushed to meet her. Another embrace, of Mother and son, I once again felt like a child, her arms protecting me, the soldier in me forgotten. My heart fluttered as she noticed my gaunt frame, stating as only a mother can, "You have lost so much weight!" All three of us embraced as some yelling co-workers in the background hailed our reunion.

Of course, Mom did not go back to work the rest of the day. My mother was one of the hardest working women I have ever known, a young girl of twelve when her father died before the depression, I wished she could have lived longer for me, but she passed in 1977.

I was twenty-two and home. During my first few days, I only had one episode of my wandering conscious returning to Vietnam. This happened one night while sleeping at Janet's parent's apartment with Janet by my side. Some couple, directly across the courtyard of this 1930s six story brick apartment building, were fighting violently, emphasized with throwing dishes or other objects against their walls, clearly within my polished enhanced combat hearing range. This exchange of loud crashing noise bolted me from my sleep, jumping/leaping onto my feet as I yelled loudly "Incoming! Incoming!" practiced now from rote memory of my year of war.

As my now open eyes scanned the perimeter to play catch up to confirm this event, again well-practiced past remembrances. How do you explain your subconscious behavior to your love ones jarred awake by an adult man screaming, but none asked.

I failed to mention in previous chapters that the majority of pilots while deployed managed to save a substantial amount of money during their tour, which I wish now I had used to buy property. A few weeks before I would depart Phan Thiet, I had seen my dream car in an advertisement in Readers Digest, an iconic magazine with varied stories each month. A 1970 Buick Grand Sport, silver with black vinyl roof and magnesium wheels, a true muscle car manufactured for a pilot. As I remember, practically every pilot bought a muscle car on their return. Bruce had a Mustang GT, others opted for Chevy Corvettes. Our reward for survival or just a gaudy car to remind us to savor this extended life some never got.

Janet had purchased this symbol, fulfilling my request at a price of $4,200.00 dollars, a substantial sum for that era, especially without financing. A secret discovered upon my return was Janet had been picking her girlfriend up at the Yonkers train station a week before my arrival only to have a Mr. Dominos Rios crash/torpedo into my dream car directly in front of the train station. Of course, Mr. Rios had no license or insurance and most likely sent Janet into panic mode with my return just weeks away. Comical now to look back because nothing really mattered then, I was alive, and it just can't get any better knowing the quick sting of death follows us all. Wise beyond my years, through experience.

Janet had rushed the repairs of the GS but like all covert operations, the unforeseen happened. I arrived one day early before the GS was completed. I didn't care, I was just ecstatic to exist.

Irony is life's equalizer; a joyful return is crushed. My wife's younger brother Jimmy had just finished his junior year of high school and at seventeen was on the cusp of his adult life. Unfortunately, he was killed by a random childlike impulse of his own adolescence. Returning with friends from a day at the beach, young Jimmy was on the subway demonstrating (to the less courageous members of this entourage) his skills of riding between the cars. These rumbling metal wagons had a small metal platform where one could negotiate moving from one car to another. What normal young man hasn't wanted to visually impress girls in his presence?

Securing himself into a position, he lifted his teenage body which would then place the crown of his head just above the train car. Regrettably, the clearance to the top of the dark tunnel is minimal; as a steel beam struck the top of Jim's skull. Janet's father, John (Jack) Panko a WW2 sailor and very well-respected NYC Police Mounted Sergeant (now memorialized with a bronze plaque in his honor at the Mounted Police Stables in Manhattan) and her mother Meriam both were utterly destroyed by this heart-rending event. With tears in his eyes and a face etched with sorrow, Jack approached me and requested if I could proceed to Bellevue Hospital morgue in NYC to identify young Jim's remains.

I had been home maybe three to four days now and was driving to Manhattan on a mission for another young man, when does it stop? At the morgue, the attendants escorted me to the viewing area: a cold sterile place of the hospital, a lifelessness room which emanates despair. The stainless-steel gurney with Jim's remains was wheeled forward. Jimmy's head had been shaven; his cold corpse lying wrapped in clear heavy plastic, his body was an achromatic color of white as his blood no longer pulsed through his youthful frame. There was a large contusion on the crown of his skull: "that's it", simply, possibly a faint scrape. A fraction of a second type impact had taken young Jim's life.

Before starting my new assignment at Ft Rucker, I attended a party of high school friends one evening in Yonkers and felt detached as they chatted and laughed, drinking legally but still acting like high school alumni, normal for them. Janet struggled; it had been three weeks since Jimmy passed, but close friends helped. I was still in Vietnam where my aviation friends were counting down their time.

After arriving at Ft Rucker and being selected to attend an eight-week course to become an Army Instructor pilot, normality began to return. I was one of five pilots who graduated MOI (Method of Instruction) course out of the original fifteen pilots, most failed out, they just had X number of flying situations and needed classroom ground instructors, their own version of a purge. I had my flying job and truly loved my position of training future Army pilots while feeling that deep concern for their own upcoming journey.

Some will berate the Armed forces and looking back during this time in our country's history, many more would choose this form of scrutiny, but for myself it was a lighthouse of direction. They took an interest and gave it a place to grow. Their perfected learning procedures of programed texts and speed-reading classes were cutting edge for this period.

How many times in all these many years when people often ask, I impulsively expound "yes — our youth with no direction should join the armed services." The young should endeavor to get out of a stale environment. They should move from bad neighborhood streets and disruptive associates who are wasting away with no goals, no drive, no hope, but whose lives and potential shrivel, undeveloped, dragging most who stay with them into only dreams of what may have been. At least those who recognize their plight may discover a much grander world where someone will give them responsibility with multiple avenues of varied educational opportunities.

"Time is a notion of man, but memories are the product of one's being."

My quote, I think. However, centuries of true scholars may have uttered such thoughts. So just in case, "Mea Culpa."

THOMAS MCGURN

DATES I WILL ALWAYS REMEMBER
OF MY YEAR IN VIETNAM:

July 20, 1969, first manned landing on the moon
October 16, 1969, NY Mets win World Series
February 24, 1970, It happened so fast
April 29, 1970, Nixon invasion of Cambodia
May 4, 1970, Kent State killing

Post War

An interesting encounter many many years later, but a preeminent example of the circle of life experience, my opinion again drawn from a Disney inspiration, I'm an Army pilot not Nostradamus.

Sometime in the Autumn of the late 1980s I was in the NYANG as an instructor pilot, a weekend soldier being assigned to conduct/lead a two-ship flyby for a memorial dedicated to the nurses of the Vietnam War/Conflict. This location was Ladson Park in Westchester County, NY. A beautiful early fall setting at the end of a winding trail through the changing foliage of the woods. The memorial figures were life size bronze sculptures on a natural stone base looking out onto a reservoir lake. Truly fitting to stimulate that contemplative thought process on those who sacrificed their lives, their existence, for fellow Americans.

Immediately after our low and slow formation pass of this heart moving memorial, we executed our pre-planned landing on the Park grounds for a static helicopter display. Our machines now being a visual image of the Vietnam war, two UH-1H Hueys'. This war wagon whose sound and silhouette need no explanation, they are probably the best memory enhancer of that turbulent past. We immediately became an attraction as families with young children surrounded the aircraft. Many were Vietnam Veterans, their emotions stirred now, trying to explain their experiences to their precious loved ones who have heard the stories, but now with this static relic present on this day of the memorial they can touch and feel while climbing onboard, their senses alerted through past experiences. Many will politely ask if they can sit inside (as if they didn't earn or deserve that privilege), now reliving their lost year of youth. You can clearly observe these Veterans as their posture transforms them, that some sense of fulfillment returns to their souls. I loved these static displays, it was cathartic.

Approaching from the rear of the aircraft on the expansive green grass lawn was an older gentleman, casually dressed. His erectness was obvious with a gray head of hair and the walk of a Veteran. Unbelievable! As I recognized those distinctive eyebrows, General Westmorland approached, as civilian General Westmorland. A quick salute from me to an icon of my past. The general is humble, not the TV presence I had seen so many times, a man whose mission and time had passed. Just another old soldier saying hello to soldiers, a true sign of an honorable man of character and presence. I cannot recall the brief conversation that took place, but I felt no distain for the past. I understood a general's decisions most likely reflected the restraints imposed by his supervisors, the President of the United States and Secretary of Defense. How many times have all of us experienced this small world factor? To this day, this encounter enforces in my mind the humility of our Military.

Ladson Park

Ladson Park

In 1995, I would transition to UH-60 Blackhawks, leaving my beloved UH1-H workhorse forever, missing that sound and total confidence of comfortability. In 2002, just after the 911 attack, I would be assigned to participate in moving three Al Qaeda prisoners in late October or the first week in November to a federal prison in Pennsylvania. It was a Sunday morning, clear but very turbulent, as we landed three UH-60A

Blackhawks on the NYC Downtown Heliport (old Wall Street). This could have been a scene from a high-end action movie. Airspace completely shut down except for us. NYPD, Coast Guard, US Marshalls everywhere, even closing the FDR drive, a visual indication of the importance placed on this mission. NYPD emergency officers with automatic weapons were now covering down. The harbor was filled with Police boats and Coast Guard, to think this was the same width of sea water that General George Washington used to escape the British Army from Brooklyn Heights during the Revolutionary War in August of 1776.

As the prisoners arrived, their heads were hooded in black to prevent any errant observations which they might make of the recent destruction during the 911 attack, although that acrid smell which I would describe as electrical wire burning still filled lower Manhattan from the debris field formerly called the World Trade Towers. Particularly disturbing were the barges now being loaded with the twisted and mangled steel beams/girders lined up inside the heliport platform, instantaneous reminders of all the lost innocents. I considered this mission an honor to transport these despots and one of the most memorable moments of my aviation career.

In September of 2004, I had been chosen to go to the Iraq war, not my choice but this was an obligation I had accepted; it echoed my decision in 1968. Personal note "never volunteer", but never hide. How fortunate to survive, how fortunate to experience a full life. For all those who have now read this old soldier's reminiscences, remember that the next day is a new day, a new chance, always anticipate and embrace those possibilities.

The Special Ones

WRIGHT JOHN PAUL

Name: WO1 John Paul Wright
Status: Killed In Action from an incident on 10/28/1969 while performing the duty of Pilot.
Age at death: 20.0
Date of Birth: 11/11/1949
Home City: Aurora, CO
Service: AV branch of the reserve component of the U.S. Army.
Unit: 192 AHC
Major organization: 1st Aviation Brigade
Flight class: 69-7
Service: AV branch of the U.S. Army.
The Wall location: 16W-001
Short Summary: Gunship flew into a mountain at night while covering rangers with Thomas Campbell.
Aircraft: UH-1C tail number 66-15075
Call sign: Tiger Sharks
Service number: W3164496
Country: South Vietnam
MOS: 100B = Utility/Observation Helicopter Pilot

Primary cause: Hostile Fire

Major attributing cause: aircraft connected not at sea

Compliment cause: fire or burns

Vehicle involved: helicopter

Position in vehicle: pilot

Started Tour: 05/20/1969

"Official" listing: helicopter air casualty—pilot

Location: Ninh Thuan Province II Corps.

Reason: aircraft lost or crashed

Casualty type: Hostile—killed

single male U.S. citizen

Race: Caucasian

Religion: Methodist (Evangelical United Brethren)

The following information secondary, but may help in explaining this incident.

Category of casualty as defined by the Army: battle dead Category of personnel: active duty Army Military class: warrant officer

This record was last updated on 07/25/1998

HELICOPTER UH-1C 66-15075

Information on U.S. Army helicopter UH-1C tail number 66-15075

The Army purchased this helicopter 0367

Total flight hours at this point: 00001335

Date: 10/28/1969

Incident number: 69102890.KIA

CHECK RIDE

```
Unit: 192 AHC
This was a Combat incident. This helicopter was LOSS
TO INVENTORY for Close Air Support
Unknown this helicopter was Unknown at UNK feet and
UNK knots.
South Vietnam
UTM grid coordinates: ZT259230
Helicopter took 1 hits from:
causing a Fire.
Systems damaged were: PERSONNEL
Casualties = 04 DOI . .
The helicopter Crashed. Aircraft Destroyed.
Both mission and flight capability were terminated.
Burned
```

CREW MEMBERS:

```
P   CPT CAMPBELL THOMAS EUGENE KIA
P   WO1 WRIGHT JOHN PAUL KIA
CE  SP5 LOWREY CHUBBY DEAN KIA
G   PFC TORRES EZEQUIEL JR KIA
```

CW2 Norman E. Niswonger was a potential VHPA member who died after his tour in Vietnam on 12/15/1979 at the age of 31 from Tail boom broke off in flight due to stress cracks.

```
Dayton, OH
Flight Class 69-33/69-31
Date of Birth 11/12/1948
Served in the U.S. Army
```

THOMAS MCGURN

Served in Vietnam with 192 AHC in 69-70

This information was provided by Roger Mitchell, 7th & 6th Reunion, Eric B Williams

More detail on this person: Pilot Normam Niswonger, 31, of Piqua, was killed Saturday (12/15/79) when the shaft driving the tail rotor of his helicopter snapped and his craft plunged to the ground out of control. NISWONGER was president of Helitractor Corp., which leased the copter to WHIO-TV of Dayton. NTSB accident report is at http://www.ntsb.gov/recs/letters/1980/A80_9_10.pdf The tail boom separated from the helicopter due to stress cracks.

This information was last updated 02/13/2010

CHECK RIDE

WO Rocky D. Armstead
Born: March 30, 1946
Died: October 5, 1969

About the Author

Tom McGurn was born and raised in Yonkers New York and attended Sacred Heart High School.

He is a combat veteran of two wars and has served his nation for forty years as an Army Aviator. Tom served in Vietnam from 1969 to 1970. After arriving home from Vietnam, Tom attended and graduated the US Army flight instructor pilot course and immediately started to train young pilots. He attended numerous US Army sponsored courses including the Electronic Warfare course. He served in the Iraq War (2004-2005) where he was assigned to the G3 Air section of the Division TOC to coordinate the 42nd Divisions air movement assets. Based on his training, he was the Tactical Operation Officer.

In addition to his active duty, Tom was in the New York Army National Guard. He is a retired supervisor of the Westchester County Police Department where during this dual career path he was a Detective, and Detective Sergeant in Narcotics along with many varied police assignments.

Chief McGurn retired in 2008 from the United States Army as a Chief Warrant Officer Four.

www.ingramcontent.com/pod-product-compliance
Lightning Source LLC
Chambersburg PA
CBHW030318100526
44592CB00010B/482